Sociolinguistics and Mobile Co

Edinburgh Sociolinguistics

Series Editors:
Paul Kerswill (Lancaster University)
Joan Swann (Open University)

Volumes available in the series:
Paul Baker, *Sociolinguistics and Corpus Linguistics*
Ana Deumert, *Sociolinguistics and Mobile Communication*
Scott F. Kiesling, *Linguistic Variation and Change*
Theresa Lillis, *The Sociolinguistics of Writing*

Forthcoming titles include:
Kevin Watson, *English Sociophonetics*

Visit the Edinburgh Sociolinguistics website at www.euppublishing.com/series/edss

Sociolinguistics and Mobile Communication

Ana Deumert

EDINBURGH
University Press

Edinburgh University Press Ltd
The Tun - Holyrood Road
12(2f) Jackson's Entry
Edinburgh EH8 8PJ
www.euppublishing.com

Typeset in 10/12pt Adobe Garamond by
Servis Filmsetting Ltd, Stockport, Cheshire,
and printed and bound in the United States
of America

A CIP record for this book is available from the British Library

ISBN 978 0 7486 5573 1 (hardback)
ISBN 978 0 7486 5575 5 (webready PDF)
ISBN 978 0 7486 5574 8 (paperback)
ISBN 978 0 7486 5577 9 (epub)

Contents

Figures

Tables

Acknowledgments

Completing this book took longer than planned. The text kept growing and changing, developed a life of its own, and took me to places, theoretically as well as virtually, which I hadn't anticipated. Like any book, it was an intellectual journey, and I am grateful to those who helped and supported me on the way.

A big 'thank you' goes to all my students who were involved in data collection during the last few years, and kept me on my toes with difficult questions. In particular I would like to thank: Yolandi Klein, Oscar Sibabalwe Masinyana and Zolani Kupe (South Africa), Beatrice Bruku (Ghana), Tolu Odebunmi (Nigeria), Epiphane Sedji (Togo) and Stéphane Pepe (Côte d'Ivoire). I would also like to thank all those who spoke to us, shared their texts with us and allowed me permission to reproduce them.

Very special thanks go to Don Kulick for introducing me to Derrida, commenting on each chapter with rigor and passion, and generously engaging with my thinking in ways that allowed me to refine and develop it. Heartfelt thanks are due to Chris Stroud, who not only was my collaborator on the original research project but read the manuscript carefully and critically, and was an invaluable sounding board over coffee and drinks. I am indebted to Cécile Vigouroux for asking me difficult and important questions about what it means to do digital ethnography; to Marion Walton, who taught me all I know about the technical side of digital media and always reminds me about inequalities of access; to John Singler and Margot van den Berg and our discussions – online and offline – about texting in Africa; to Jan Blommaert, who encouraged me when I had just started; to Rajend Mesthrie whose collegial support has been invaluable over the years; to Joan Swann and Paul Kerswill, who have been incredibly wonderful and patient editors; and to my colleagues, in Cape Town and elsewhere, who gave me opportunities to present my work, and who were there to listen, to talk, to discuss and to debate. The biggest thanks of all go to my husband, friend, partner and intellectual soul mate Nkululeko Mabandla. He was there every step of the way, read the manuscript with a beady eye, and never let me get away with shoddy thinking. *Ndiyabulela Bhele*!

The work reported was financially supported by the National Research Foundation (South Africa), SANPAD (South African Netherlands Project for Alternatives in Development), the Shuttleworth Foundation (South Africa) and the University of Cape Town.

For assistance with checking the different scripts in the text I would like to thank Rajend Mesthrie (Devanagari), Huamei Han and Xuan Wang (Chinese languages) and Shamil Jeppie (Arabic).

Chapter 1

Media sociolinguistics

INTRODUCTION: SOCIOLINGUISTICS AND THE MEDIA

In 2012, while doing fieldwork in the rural areas of the Eastern Cape, South Africa, I interviewed a 75-year-old subsistence farmer. The interview, which took place on an ordinary Saturday morning, was repeatedly disrupted by the ringing of her mobile phone: her husband called from Johannesburg, her son from Mthatha, her sister from Cape Town, and a church member from the neighboring village. Although Ma Goxo had limited schooling and struggled with the small keypad of the phone, she had found ways to use text messages too: her grandchildren would type them and read them out to her. When I asked whether she could imagine life without her mobile phone she looked at me in disbelief, then exclaimed in isiXhosa, *soze*! ('never'). This sentiment was echoed by many others. People from a wide range of social, educational, geographic and linguistic backgrounds have invested their often meager resources in communication technologies and have become regular media users.

This book discusses people's engagement with communication technologies – mobile phones and computers – from the broad perspective of sociolinguistics, an interdisciplinary and empirically oriented field of study, which is concerned with understanding language use in its social and cultural context. The title of this chapter, 'Media Sociolinguistics', positions the discussion within a larger interdisciplinary tradition of research on media, language and society. This research involves not only linguists, but also anthropologists, sociologists and media scholars, whose interest is to describe and analyze the role of media, old and new, and the way they co-exist and interact, in social life.

Sociolinguists are increasingly exploring media and, especially, digital communication as an area of study. Media sociolinguistics allows us fresh perspectives on topics that are central to the discipline: multilingualism and linguistic diversity, language variation and change, style and register, language and identity, language ideologies, interactional linguistics, and language and globalization (Thurlow and Mroczek 2011; Seargeant and Tagg 2014). Digital language also provides us with new perspectives on writing and challenges beliefs about what writing looks like, or should look like. In addition, digital practices are relevant for the study of multimodality, that is, the ways in which language and other semiotic resources (images, layout, sound etc.) work together in the creation of meaning. The following chapters emphasize the interdisciplinary nature of sociolinguistics and introduce readers to the richness of the work that has been carried out globally. Theoretically the chapters draw on sociolinguistic key concepts such as ideology, power and discourse (especially Chapters 2–5), and explore the performative and experiential aspects of digital interactions, focusing on style, voice, performance and performativity, play and art as well as sociability (especially Chapters 6–8). This introduction locates the book within broader debates about the 'newness' of digital media, and proposes three central themes for the analysis: mobility, creativity and inequality.

THE MORE THINGS CHANGE *OR* MEET THE ANCESTORS

In the mid-nineteenth century, the French critic Jean-Baptiste Alphonse Karr (1808–90) wrote *plus ça change, plus c'est la même chose*, usually translated into English as 'the more things change, the more they stay the same'. When I started working on digital technologies and the sociolinguistic practices they enable, I kept wondering about the question of novelty: do *new* technologies create fundamentally *new* practices? And do we need equally *new* theories to understand what is happening? Do we need a sociolinguistics 2.0?

Like work on globalization, the study of new media has often been framed in terms of rupture or revolution. Arjun Appadurai (1996: 3), for example, speaks about 'a dramatic and unprecedented break between past and present', and Manuel Castells (1996: 31) maintains that communication technologies have brought about a fundamental and indeed epochal shift in 'the way we are born, we live, we sleep, we produce, we consume, we dream, we fight, or we die'. With respect to language and communication, Gunther Kress (2010: 7) suggested that '[we] do not yet have a theory which allows us to understand and account for the world of communication as it is now'. This is echoed by David Barton and Carmen Lee (2013: 16), who argue that '[b]asic linguistic concepts are changing their meaning and a new set of concepts is needed'. Taken together these discussions suggest that we cannot carry on with 'business as usual', but need new theories and concepts if we want to understand and explain the practices we observe online.

In this book I take a somewhat different view and argue that we don't need a fundamentally new type of sociolinguistics (or sociology, or anthropology) to describe and understand the practices we see online. Instead I take my cue from the work of A.L. Becker (1995: 406), and suggest that we should bring back *the ancestors*, 'who still have a lot to teach us if we can relearn how to read them'. There have been many linguists (and social scientists) in the past who have thought carefully about language, and although theirs was a world without computers and mobile phones, their writings still have a lot to teach us. While Becker's attention to ancestral voices was inspired by the Javanese tradition of *jarwa dhosok* (literally 'pushing old language', that is, the exegesis of old texts in traditional shadow play), a turn toward ancestors also has a strongly African feel to it: ancestors don't die, they are always present, as intermediaries, as inspiration and as those who guide us. And we ignore them at our own peril. Throughout the book I will argue that as linguists we have inherited an immensely rich intellectual tradition, and ways of talking about language and social context. Just because certain views were at the margins of the discipline in the past, this does not mean that they weren't there to start with, and that new technologies have put us at ground zero.

Not all of the ancestors, whom I discuss in the chapters to come, are equally well known, and not all of them are linguists. While the work of the Russian language philosopher Mikhail Bakhtin (1895–1975), as well as that of the American sociologist Erving Goffman (1922–82), are well established in contemporary sociolinguistics and new media studies, the same is not necessarily the case for some of the other ancestors: Wilhelm von Humboldt (1767–1835), Edward Sapir (1884–1939), Roman Jakobson (1896–1982), Georg Simmel (1858–1918), and also – somewhat more contemporary – Roland Barthes (1915–80) and Jacques Derrida (1930–2004). Some are linguists in a broad socio-cultural sense (Bakhtin, Humboldt, Sapir, Jakobson), others are sociologists (Simmel, Goffman) or philosophers (Barthes, Derrida).

Why these and not others? Partly this is, of course, personal preference and reading habit. However, I also argue that their work provides us with a way of thinking about language and sign-making that is helpful when considering digital practices. Their writings share a common thread by not focusing on language as an ordered and structured system, but conceptualizing it as fundamentally emergent, indeterminate and unpredictable, allowing ample scope for creativity, play and artful language. By drawing on their work and locating the overall discussion within the interdisciplinary history of sociolinguistics, I argue that the digital allows us to see things about language that are fundamental to its ontology and the way it works in social life. Digital communication draws our attention to the creative and aesthetic aspects of language, to its often wild and unpredictable diversity, and to the agency of speakers/writers as well as the constraints under which they operate.

However, trying to de-dramatize the digital does not mean that we are looking at the same-old-same-old, and that nothing has changed. We are certainly seeing major changes of scale. Although many of the sociolinguistic practices we witness online are not *per se* 'new' – that is, they do not include types of language that cannot be described and analyzed with existing theories (although some tweaking of these theories might be necessary) – what has changed is the frequency with which they occur, their potentially global reach, the rapid speed with which practices change and semiotic resources circulate. Moreover, the digital – just like the telegraph and the telephone in the past – provides us with new ways of interacting across time and space, and a variety of media forms (writing, video, photography) that used to be available to only a few have been democratized. They have become part of the everyday to such an extent that today, as never before, many of us live with technology (see Chapter 2; but also Chapter 3 on the digital divide and differences in access). This has unsettled old dichotomies such as author/audience or amateur/professional, and led to the formation of new sociolinguistic practices.

At the same time, there are also important continuities and as we engage with 'new' media we often draw on our experiences with 'old' media, that is, we tend to 'remediate' familiar practices rather than to develop something entirely new and unprecedented (Bolter and Grusin 1999). Consider, for example, the above-mentioned author/audience dichotomy, which has been a salient feature of many traditional media, but was also always porous. Letters to the editor are an example of user-generated content before the digital, and alternative media have always tried to overcome the strict division between author and audience, and encouraged audience self-presentation.

MOBILITY, CREATIVITY AND INEQUALITY

In thinking about new media from the broad and interdisciplinary perspective of sociolinguistics, I draw on three reoccurring themes or topics: mobility, creativity and inequality. In this section I briefly discuss them one by one; I reflect on them throughout the text and revisit them in the conclusion.

The theme of mobility is foregrounded in the title of this book: *Sociolinguistics and Mobile Communication*. Why 'mobile' and not, for example, 'digital'? The latter seems to be developing into a fairly unmarked option in sociolinguistics and the social sciences: *Digital Anthropology* (Horst and Miller 2012), *Understanding Digital Literacies* (Jones and Hafner 2012), *Digital Discourse* (Thurlow and Mroczek 2011), *Born Digital* (Palfrey and Gasser 2011), *Personal Connections in the Digital Age* (Baym 2010), *Digital Culture* (Creeber and Martin 2008), *Digital Borderlands* (Fornäs et al. 2002) and so forth. Yet there is also variation: an important journal in the field is called the *Journal of Computer-Mediated Communication*; Jean Aitchinson and Diana Lewis call their collection of papers *New Media Language* (2003), David Crystal (2011) uses *Internet Linguistics*, David Barton and Carmen Lee (2013) have titled their recent book *Language Online*, Phillip Seargeant and Caroline Tagg speak about *social media*, and the idea of *networked language* has also been proposed (Androutsopoulos 2013). These terms, although different in focus and emphasis, often serve as placeholders for each other, and are used interchangeably (a practice that I also follow). However, they are not full synonyms and they emphasize different aspects of the medium, such as the manner in which the technology works (*digital* as opposed to *analog*), the tools that are used (*computer-mediated*), the relative newness of these media (*new media*), a contrast to the *offline* world (*online*), the use of a particular platform (the *internet*), and the location of people in local and global networks of communication (*social, networked*). The adjective 'mobile' also occurs in the literature. This is usually with respect to 'mobile phones' or 'mobile internet', emphasizing a mode of access that is dominant globally, and is becoming ever more pervasive with smartphones and tablets (Hjorth et al. 2012; see Chapter 3).

My own experience of digital engagement can serve as an illustration of the historical development toward mobility. I first connected to the internet in the early 1990s in a small student flat in Germany. There were cables everywhere and finally a friend exclaimed excitedly: 'It's working!!' In those days the so-called personal computer (PC) was still a heavy, unattractive, grey and clumsy box that could not easily be moved: it was a *desktop*, meant to be kept on a desk (for a trip down memory lane, see www.binarydinosaurs.co.uk). Today, I still have a desktop in my office, but more commonly I use my *laptop*: an object that is tied semantically no longer to another object, but to my body (my lap). My tablet computer and smartphone allow me even greater mobility. I might be responding to a colleague's email while traveling, text a friend about dinner while rushing from one classroom to the next, or post a Facebook status update while working out at the gym. My own experiences are mirrored by global sales statistics: consumers are consistently opting for those technologies that allow them a maximum of personal mobility, away from plugs and power points.

Although this sense of personal, physical mobility is important, I also use 'mobile' in a broader sense in this book, that is, as a way of thinking about language and semiotic resources more generally. In this I follow recent work on the sociolinguistics of globalization (Blommaert 2010, 2013; Pennycook 2012; Blommaert and Rampton 2011) and the 'mobilities turn' in sociology (Urry 2007; Cresswell 2012a, 2012b). Thus, mobility affects not only people, but also languages and their relationships to each other, as well as texts and symbolic resources, and the ways in which they are understood by diverse audiences, located in multiple contexts of media consumption.

The second theme is that of creativity, a topic that is receiving growing attention in sociolinguistics more broadly (Jones 2012; Swann et al. 2011; Alim 2011; Maybin and Swann 2007; Carter 2004; Sherzer 2002; Crystal 1998). In her book *Cyberpl@y*, Brenda Danet (2001) has argued that creativity, art and play are central to digital engagement. Early examples of this are found on the interactive bulletin board systems of the 1970s and 1980s and on Internet Relay Chat in the 1990s: writers play around with language, use typography in new ways, make puns and joke (Baym 1995). From the early 2000s onward, Web 2.0 applications – Facebook, Flickr, YouTube, Twitter and many others – have embedded creativity, that is, the production of content rather than merely its consumption, firmly in digital architectures and the discourses of online engagement (Jenkins 2006).

Examples (1)–(3) are birthday greetings that were posted on the Facebook page of Kushie, a female South African student who had just turned nineteen (in 2011). Such language practices challenge assumptions not only of what written language *should* look like, but also of what it *can* look like. Is it a case of anything goes? Can writers invent new written forms on a whim, or are there limits to their creativity?

(1) happy borndy Kushie nj0y t mwha

(2) happy bdae to u gal ts so0 0fficial mndae we gttn drunk!!! Eeenjooy ur dae mrw mcwah

(3) Hhhhhaaaapppppyyyyy Bbbbdddaaaayyyy!!!!!!!!!!!!! enjoy it 2 da fullest!! love u2 bits!

One of the reasons we struggle to deal with creativity is that in sociolinguistics – as well as in the social sciences more generally – the focus has historically been on ordered patterns and conventions, rather than the unexpected, that which is strange and surprising. While fully aware that speakers and writers are creative – that they pun and subvert, play around and invent new forms – linguists have generally pushed this knowledge to the margins of the discipline. We certainly enjoy reading books about language play, but we don't usually teach it to our students: play and creativity might make for a good and engaging lecture, but the real business of sociolinguistics – and even more so linguistics – appears to be elsewhere. When I took a self-critical look at *Introducing Sociolinguistics*, a textbook that I co-authored and that appeared in a second revised edition in 2009, I realized that there are no entries in the index for creativity, play or verbal art (Mesthrie et al. 2009). In this book I wish to step away from this tradition and position creativity as fundamental not only to digital language, but to language in general.

The third theme of the book is inequality, and the discussion reflects my own positionality as a linguist located on the African continent, where digital engagement is growing but still limited and restricted when compared to the global North. In approaching digital inequality I follow the anthropologists Heather Horst and Daniel Miller (2012: 25), who have argued that when writing about digital interactions and practices, we need to keep their *materiality* in mind: 'Materiality . . . is bedrock for digital anthropology.' Technologies always produce a material footprint: factories that produce them, billboards that advertise them and shops that sell them. The content that users produce has its own materiality too. It is stored in bits and bytes, becomes visible on screens as text or image, can be printed on paper or copied by hand. And finally there is the materiality of context, the very real bodies of people and the socio-economic constraints within which they operate, especially their ability, or lack thereof, to invest in technology and to use it.

Although sociolinguistics has been shaped by a long-standing commitment to questions of inequality and justice (Bucholtz and Hall 2008; Zentella 1997), sociolinguistic studies of new media have so far failed to articulate a meaningful global perspective (a similar observation has been made about media studies more generally; see Curran and Park 2000). For example, Naomi Baron's (2008) book *Always On*, which is subtitled *Language in an Online and Mobile World* (my emphasis), focuses almost exclusively on experiences in the United States, with just three pages (out of almost three hundred) providing what she calls a 'Cross-Cultural Sampler'. Too often global differences are silenced in sociolinguistic discussions of online practices: while there might be a passing acknowledgment that not everyone in the world is connected to digital technologies and the internet, the actual patterns of global access are rarely discussed, and generalizations about digital practices are often based on data collected in contexts of socio-economic affluence. The situation is somewhat different in anthropology, where there exists a strong tradition of media research in societies where resources are scarce (e.g. Horst and Miller 2006; De Bruijn et al. 2009; see also the papers edited by Ling and Horst 2011). Anthropological work is important to consider and I draw on it throughout the book. However, for a sociolinguist, the anthropologist's lack of attention to the micro-details of language and communication practices can, at times, be a source of frustration.

Throughout the discussion I also draw on my own sociolinguistic material, which was collected in South Africa between 2010 and 2013, and which focuses on Afrikaans, English and isiXhosa.[1] South Africa, a so-called emerging economy, is a socio-economically highly polarized society, and this allows one to look at digital engagement from the perspective of relative affluence as well as intense scarcity (Chapter 3).

OVERVIEW OF CHAPTERS

The chapters in this book approach the topic of mobile communication from different angles, drawing on a range of sociolinguistic concepts, topics and approaches.

Chapter 2 considers Erving Goffman's notion of the interaction order from the perspective of mediation. Digital media differ from body-to-body communication in important ways. They allow us to interact across time/space and to archive our interactions.[2] This facilitates the mobility of people and texts, and allows for new forms of creativity. In theorizing mobility, I draw on work in sociolinguistics as well as sociology. The section on creativity brings in two ancestors: Humboldt and Sapir, both of whom emphasize the importance of sign-making as creative construction, rather than rule-following. The chapter concludes with a discussion of methodological and ethical issues.

Chapter 3 focuses on the materiality and political economy of digital communication. Communication technologies enable us to do things, but their very materiality also limits the actions we can perform. The chapter provides a global overview of digital access and usage, and explores practices of digital engagement in contexts of affluence and poverty, not only distinguishing the haves from the have-nots, but also taking a close look at the have-less people. The data that is presented in this book needs to be understood within the context of these material conditions.

Chapter 4 continues the discussion and looks at the increasing geographical diversification of internet users and the growth of digital multilingualism. The chapter adopts the methodological and theoretical lens of linguistic landscaping and asks how this can be extended to the study of virtual spaces, using Wikipedia as a case study. The chapter shows that although online spaces hold the promise of empowering marginalized languages, they can also perpetuate marginalization and create public representations of language that are deeply problematic.

Chapter 5 focuses on the multimodal affordances of, especially, YouTube. On YouTube, audiences actively engage with what they see: they might comment, express their like/dislike, and produce mash-ups or remixes as response videos. The central theoretical notion introduced in this chapter is intertextuality, that is, the ways in which texts always relate back to earlier texts, and gain new meanings as they are being recontextualized. Drawing on the language-philosophical work of Derrida, the chapter also traces the limits of intentionality, and the ways in which meanings can multiply well beyond the intentions of the author.

Multimodal affordances notwithstanding, written language remains – for the time being – central to online interaction. Chapter 6 and Chapter 7 look at writing in weakly regulated contexts, that is, digital spaces where standard norms are relaxed and experimentation is encouraged. Chapter 6 starts by considering the relationship between speech and writing. I argue that reflexivity is an important aspect of writing, and proceed to approach digital writing from the perspective of verbal art and performance. The chapter introduces and illustrates Bakhtinian concepts such as heteroglossia and stylization, as well as Richard Bauman's work on performance and Judith Butler's idea of performativity. Chapter 7 continues the argument of digital writing as artful performance by looking more closely at orthography and typography, that is, the visual aesthetics of writing. Drawing on Roman Jakobson's work on poetic language and locating it within Jakobson's own poetic practice and the artistic context of the time, I argue that digital writing recontextualizes poetic strategies of early twentieth-century avant-garde poetry.

Chapter 8 moves from a focus on form to the things people do online: how a sense of community is built, and how identities are created and transcended. The chapter identifies two main types of online social engagement, which – drawing on Roland Barthes – are referred to as *plaisir* and *jouissance*. Much of what happens online is *plaisir*, reflecting an enjoyment of togetherness that does not challenge the status quo and creates new spaces of intimacy. However, the digital is also a place of transgression and subversion, creating the more ambiguous enjoyment of *jouissance*, anarchic and often discomforting. The chapter relates these two types of digital interaction and enjoyment to Simmel's work on sociability and Bakhtin's notion of the 'carnivalesque'.

Chapter 9, the conclusion, then revisits the three main themes of the book: mobility, creativity and inequality.

A note on typographic conventions in the text: italics are used for non-English material, in-text examples, book titles as well as for emphasis. Each chapter, with the exception of Chapters 4 and 9, has an anonymized Twitter status update as a motto.

All links cited in the text were live in April 2013 when I completed and submitted the manuscript. However, data permanence is an issue for online research (see Chapter 2). The popular remix practices of, especially, YouTube (discussed in Chapter 4) mean that accounts can be terminated 'due to multiple third party notifications of copyright infringement', and that video links suddenly become unavailable. Although we can keep personal copies of the files, they are no longer available to readers.

NOTES

1. I use emic African terminology when referring to groups of people (*amaXhosa*, 'the Xhosa people') and their languages (*isiXhosa*, 'the Xhosa language').
2. I follow Leopoldina Fortunati (2005: 53) in using 'body-to-body' rather than 'face-to-face' because 'it expresses more accurately all the richness of communication between co-present individuals'.

Chapter 2

Mapping the terrain

> Twitter, alive with possibilities . . .
>
> Twitter 2013

INTRODUCTION: A NEW INTERACTION ORDER?

There exists a deep-seated belief that 'good' communication is 'soul-to-soul, among embodied live people', and that mediated engagement is potentially problematic, impoverished and prone to misunderstandings (Peters 1999: 47). The French philosopher Jacques Derrida ([1972] 1988) called this assumption the metaphysics of presence; that is, a long-standing belief in Western philosophy, going back to Socrates, that physical co-presence is ontologically privileged and superior to forms of mediated presence. A focus on physical co-presence has also informed work in sociolinguistics, and underpins Erving Goffman's concept of the interaction order; that is, the analysis of the shared practices and understandings that guide everyday interactions among people, and that shape social life. Goffman – writing at a time when computers were not yet widespread and mobile phones only seen in the hands of wealthy people – limited the interaction order to situations where 'two or more individuals are *physically* in one another's response presence', and he suggested that phone conversations and letters are 'reduced versions of the *primordial real thing*' (1983: 2, my emphasis). Today, with computers and mobile phones forming an integral part of our lives, used for even the most intimate interactions, researchers do not necessarily agree with his view, and media scholars have argued that the idea of the interaction order applies equally to mediated interactions (Rettie 2009).

This chapter discusses the core characteristics of the mediated interaction order. The first section looks at the ways in which mediated interactions differ from body-to-body interactions: mediated interactions reorganize time-space relations between participants and transcend the ephemerality of spoken words. By enabling interaction in the absence of physical co-presence, new media allow us to experience mobility in new ways. We can remain connected to, and in conversation with, known and unknown others even though we are miles apart. Mobility refers not only to people moving. It also includes the movement of ideas and semiotic resources, and this mobility too is facilitated by contemporary media, especially by the fact that texts can be saved, edited and reproduced easily. The theme of mobility is explored in the second section of this chapter.

In addition to mobility, creative practices have been noted as a key feature of new media engagement. As discussed in the previous chapter, people do not only consume content, but also produce content for others. This means that we need to think carefully about creativity, within sociolinguistics and beyond. In the third section of this chapter, I draw on the work of Wilhelm von Humboldt and Edward Sapir – written in the early nineteenth and early twentieth century respectively – as a theoretical starting point for thinking about creativity, language and, indeed, sign-making more generally. And finally, like all social research, the sociolinguistic study of mobile communication raises methodological and ethical issues. The final section of the chapter discusses various approaches to online methodology and provides a brief ethical guide. The overall discussion in this chapter – as well as in the following chapter – is broad and deliberately interdisciplinary, drawing not only on sociolinguistics but also on social theory and philosophy in exploring the interaction order in the age of technology.

ACROSS SPACE AND TIME: THE VIRTUAL AND ITS ARCHIVE

The British sociologist Anthony Giddens (1990) describes time-space distantiation as one of the characteristic features of modernity. Social relations are no longer restricted to conditions of physical co-presence, but they are increasingly lifted out of 'local contexts of interaction' and stretched across time and space (p. 21). Communication technologies are central to the experience of this reorganization of time-space. We might hear about an earthquake on the other side of the world within seconds on TV and witness how it unfolds in real time. We might receive the news of a friend's first child or his most recent break-up in a letter, text message or email, and discuss the outcome of the latest Academy awards with people from across the world on Twitter. Being engaged in such *translocal* contexts and interactions produces overlapping communicative frames. We are both here and there, rooted in a physical context, but also elsewhere with others. This dual orientation supports sociolinguistic practices that Marco Jacquemet (2005) calls transidiomatic; that is, practices, that transcend traditional local norms and combine translocal – often global – linguistic and semiotic resources (see also Meyrowitz 1985, and Chapters 5, 6 and 7 for a detailed discussion and examples).

Translocality belongs to the realm of what philosophers call the virtual. We are interacting with people, objects and events that are *physically absent* as if they were *present*. It is important to remember that such experiences of absent presence are not new and human societies have attended to the virtual, the possibility of imagined presence, long before the introduction of computers and mobile phones. Praying to deities, talking to the dead, dreams, mirrors, optical illusions, the monetary system, as well as books, letters and telephone conversations – in all these cases we experience, quite habitually, a presence in the face of physical absence.

In everyday language, the virtual is often contrasted with the *real*. 'The server is lagging again! Time for real life!' might be the closing line in a chat conversation before logging off. But what is real? Philosophers have always emphasized that the virtual has its own reality, and that it should be contrasted not with the real, but with the *actual*, that is, the *concrete* (Deleuze [1977] 2002; Baudrillard 1983). I can, for example raise a virtual glass to someone to celebrate. The fact that there is no concrete, physical glass in my hand does not make the act and experience of celebrating any less meaningful, and less real, to me or those around me.

In the context of digital communication, the reality and persistent ambiguity of virtual experiences were illustrated vividly in Julian Dibbell's (1993) classic article 'A Rape in Cyberspace'. Dibell describes 'a rape' that occurred in an early virtual environment called LambdaMOO, a text-based platform to which multiple users can connect at the same time, and where they can interact with one another. Individual users are represented by avatars, the user's virtual alter ego.[1] The episode told by Dibbell involves an avatar called Mr Bungle, who created a program that allowed him to take over the avatars of other users, and to make them perform certain acts – in this case sexual practices – against their will. Participants would suddenly see on their screen that their alter egos, whom they could no longer control, were engaged in humiliating acts:

> And thus a woman in Haverford, Pennsylvania, whose account . . . attached her to a character she called Moondreamer, was given the unasked-for opportunity to read the words *As if against her will, Moondreamer jabs a steak knife up her ass, causing immense joy. You hear Mr. Bungle laughing evilly in the distance.* (Dibbell 1993)

All of this was purely textual and occurred long before avatars could move cartoon-like through virtual worlds. Yet, even though these were just words on a screen, those involved in the incident felt deeply violated by the events and struggled to deal with a situation where they experienced emotional trauma, but were also aware that their actual bodies were still intact and that the experience was about representation, rather than 'the really real', the concrete. This ambiguity is visible in a post by exu, a trickster spirit of indeterminate gender, who struggles to articulate a coherent response to what happened. He/she moves from calling for Mr Bungle's 'virtual castration', expressing rage and unfiltered anger at what happened, to accusing him of little more than a breach of 'civility', something people just don't do on LambdaMOO.

> I'm not sure what I'm calling for. Virtual castration, if I could manage it. Mostly, [this type of thing] doesn't happen here. Mostly, perhaps I thought it wouldn't happen to me. Mostly, I trust people to conduct themselves with some veneer of civility. Mostly, I want his ass. (Dibbell 1993)

In his book *The Virtual* (2003), Rob Shields draws on Plato's notion of metaxis (or *metaxy*, meaning the 'in-between' or 'middle ground') to explain our ability to integrate the virtual into our everyday lives and to accept its reality, while simultaneously maintaining an awareness of its ambiguity, that is, the fact that it is different from the concrete, physical world that surrounds us. Metaxis describes a state of being in which we belong to two worlds simultaneously. It is not that we are neither here nor there, or there but not here; rather we are both here and there. The concept has been influential in drama education, and the Brazilian director Augusto Boal (1995: 43) describes it as 'the state of belonging completely and simultaneously to two different autonomous worlds'. Consider, for example, the actress Meryl Streep in the movie *The Iron Lady*. When acting she is simultaneously Margaret Thatcher and Meryl Streep; she is not one or the other. Similarly, when we communicate via telephone, email, text message or chat, we are present in a *physical, geographical place* (at the gym, at home, on a bus); yet we are also present elsewhere, in a *virtual place* with the person we are talking to. This dual presence is a fundamental characteristic of all mediated interaction, and shapes the experiential dimension of translocality and the practices that accompany it.

The feeling and experience of being together while physically distant is often accompanied by co-temporality (synchronicity), where the immediacy of response enhances the experience of shared presence. Synchronicity establishes a type of interaction that – akin to body-to-body interaction and thus to Goffman's idea of the interaction order – requires close mutual monitoring and a shared focus of attention. In asynchronous interactions, on the other hand, there is a temporal delay and the requirement for shared attention is relaxed.

Synchronicity, however, is not necessary to establish feelings of togetherness, and asynchronous communication can also work in this way. A sense of shared presence can be *brought about*, for example, by linguistic means, that is, the ways in which texts are written. Thus, an appeal to shared knowledge or the use of the present tense in Facebook status updates – 'Michael is medium-term planning!!!' – creates a sense of immediacy and intimacy, even if the update is only read days after it was posted (Page 2010). Not only can a sense of co-presence be brought about in interaction, but also synchronicity is not a stable feature of particular digital genres or applications. Rather it is the result of social norms as well as contextual factors. For example, although text messaging is usually categorized as asynchronous, expectations of an immediate response can mean that texting is *experienced* as synchronous. And sometimes chats can be asynchronous because participants might be engaged in other activities and interaction orders. In such cases, delayed responses are acceptable (for example, when chatting at work; see also the discussion of communicative affordances in Chapter 3).

Virtuality allows us to communicate not only with specific, known others in distant places, but also – via public platforms such as Twitter – with an indefinite range of potential, often unknown recipients. In contexts of physical co-presence the opportunity to address large, amorphous audiences is generally limited to newsreaders, politicians, rock stars and famous actors. In mediated interaction, the participatory structures of Web 2.0. have made it possible for everyone with an internet connection to address a crowd.

Allan Bell's (1984, 2001) work on the ways in which radio newsreaders orient their speech toward their audiences helps us to understand the sociolinguistic processes that underpin stylistic choices in mass-mediated contexts. Bell draws on speech accommodation theory, which argues that – in contexts of body-to-body interaction – speakers typically orient their language to that of their interlocutors. Bell applies this idea to mass-media interaction and shows that, even in the absence of an identifiable second-person addressee, radio newsreaders orient toward an imagined audience, and one and the same newsreader might speak quite differently depending on the audience she believes she is addressing. Bell calls this 'audience design'. Similar processes are at work in online environments. Consider, for example, Facebook or Twitter, where status updates are visible to a large group of 'friends' or 'followers'. This might include friends of friends, parents and co-workers as well as people we have never met. The fact that we know that our posts can be seen by this large audience may shape the way we write, that is, the language we choose, and how we craft our posts. And at times, we might write in ways that reflect what media scholars have called 'context collapse'. That is, we find ourselves in a situation where we address a very large number of people who are located in diverse contexts, and we are – because of the architecture of the site – not able to limit our postings to specific audiences (boyd 2002; Wesch 2009; Marwick and boyd 2011).

However, context collapse is not a general or necessary feature of digital communication. Sociolinguistic research has shown that linguistic choices are not only *responsive* to a real or imagined audience, they can also be *initiative*, that is, used to shape and (re)define the situation, to address specific groups or people, and to exclude others. In such cases people draw on what Bell calls 'referee design'; that is, they speak or write in a way that would be appreciated by those with whom they feel a sense of affinity (and who act as invisible referees of the performance; this is similar to the acts of identity model of Le Page and Tabouret-Keller 1985). In this way, digital writers are able to address particular groups within a larger, even potentially global, audience; that is, they create spaces of intimacy by writing in a particular way, or by using particular languages (Tagg and Seargeant 2014).

Body-to-body communication not only takes place in the realm of the concrete, but is also – unless we bring an audio or video recorder with us – ephemeral, present in the here and now and gone tomorrow. Virtual interactions, on the other hand, leave a material footprint, and content, once produced, can be stored, retrieved, copied and replicated well beyond its original interactional context (boyd 2010). The ability to archive allows users to look back, to remember and revisit past actions and interactions. It also allows for new forms of creativity as we can repost and remix archived texts (see Chapters 5 and 6 for examples and further discussion). The fact that messages remain available also allows others access to this data, creating new forms of surveillance and making us vulnerable. In 2013, the massive surveillance program of the US National Security Agency (NSA), leaked by former NSA contractor Edward Snowden, showed that not only can the vast amounts of online data produced by ordinary people as they go about their daily lives be collected, analyzed and stored, but this can be done without their knowing (on surveillance see also Chapter 3). Similarly, platforms such as Facebook and Google keep track of our online activities, and engage in targeted advertising on the basis of semantic analysis of our status updates or emails. In a digital world where tweets, Facebook status updates, changes to Wikipedia entries, YouTube comments and blogs are recorded and stored, the internet as a whole can be thought of as a collective, global memory place, a Foucaultian dispositive that not only assembles and stores our online practices but also contains the material artifacts, both hardware and software, that structure the way in which we perform these practices (Foucault [1977] 1980; Pentzold 2009).

While data permanence is the norm for internet data, and search engines are able to make content resurface years after its creation, not everything that has been produced remains, and the internet is also a leaky archive (SalahEldeen and Nelson 2012). How does content disappear from the internet? There are two main avenues. Firstly, websites may shut down when a company has gone bankrupt, a research project or a non-governmental organization no longer has funding, or a blog has been closed. Secondly, applications change and it can become difficult or even impossible to access texts that were created by older software or hardware. This has consequences for research practice, and Anna Everett (2002), working on African American civic society digital engagement, noted that especially non-governmental, non-profit sites can disappear very quickly.

Although archiving is common and facilitated by the materiality of digital content (stored in bits and bytes), some applications are deliberately designed *not* to archive, and to re-enact the ephemerality of the spoken word. An example of this is the imageboard 4chan, which will be discussed in detail in Chapter 8. Another example is Snapchat, founded in 2011, which allows users to interact through photos, videos and captions. The defining feature of Snapchat is that messages disappear as soon as they have been viewed, roughly after 10 seconds. There is no archiving on Snapchat and everything is intentionally and fully ephemeral.[2] However, users soon found ways to circumvent Snapchat's self-destruction algorithm, and they simply took screenshots of the text or image before it disappeared and saved those. And it did not take long for someone to come up with an application – called SnapHack– that allows users to save the image or text directly. While ephemerality has its attractions, there appears to be a human desire – 'compulsive, repetitive and nostalgic' – for the archive, for permanence (Derrida 1998: 91).

Virtuality and the ability to archive shape the mediated interaction order. They allow for the physical mobility of writers and audiences by relaxing the requirement of co-presence, and enable new forms of creative participation by facilitating the mobility of texts and signs across contexts. The next two sections consider mobility and creativity in more detail.

MOBILITY, MATERIALITY AND TETHERED SELVES

Mobility has emerged as a key concept in contemporary social theory (Urry 2000, 2007; Cresswell 2012a). The so-called mobilities paradigm considers not only the growing scale of mobility, of people, goods and ideas, since the 1950s – a development that is often summarized under the label globalization – but the very role of mobility in shaping social life. Mobility has also been emphasized in sociolinguistic theory. Jan Blommaert (2010), for example, critiques traditional views of languages and dialects as 'bounded' and 'territorialized' entities, which are conveniently depicted in linguistic atlases where each language and each dialect has its allotted and well-defined geographical place. Blommaert proposes instead a sociolinguistics of mobility, which, like the mobilities paradigm in sociology and social geography, focuses on the movement of people and semiotic resources across geographical as well as social spaces. Research in this tradition has also drawn on Steven Vertovec's (2007) notion of superdiversity; that is, the observation that diversity itself seems to be diversifying in the twenty-first century. Stable patterns of variation are fragmenting and the resulting practices – unstable and exceedingly diverse from one person to the next, from one context to the next – pose challenges for traditional, modernist sociolinguistic description (Blommaert and Rampton 2011; Blommaert 2013). Alistair Pennycook's (2012) discussion of languages, genres, discourses and images turning up in *unexpected places* similarly emphasizes the prevalence of semiotic mobility in a globalized world. Digital communication technologies are central to contemporary mobility. Texts, including video and audio material, can now be distributed to audiences at the click of a button and at great speed. Thus, in 2012, the dance moves of PSY's 'Gangnam Style' spread via YouTube across the world within weeks; a year later the 'Harlem Shake' videos spread even faster. Once we accept that mobility is widespread and not an exception, social life becomes increasingly complex and unpredictable. Shared knowledge can no longer be assumed in contexts of growing diversity, and the range of available semiotic resources – forms as well as meanings – multiplies.

Mobility needs to be thought of together with immobility, that is, people or texts who/that are limited in their movement, or whose movement is blocked or slowed down (Cresswell 2012b). Jan Blommaert (2008), for example, introduced the term 'grassroots literacies' to describe texts that are locally meaningful, but that do not travel well. When they are lifted out of the local context in which they were created, they are difficult for distant, non-local audiences to understand and interpret. I return to the limited mobility of grassroots writing in more detail in Chapter 4. People experience immobility too, among them women who fear going out alone at night, people with disabilities who cannot move freely in a city designed for the able-bodied, and those who lack the funds to travel. In such contexts digital spaces can provide virtual mobilities. Being able to 'travel through the phone' or 'travel through the computer' emerges as an important form of mobility, and creates opportunities for non-local interactions despite physical immobility. Jenna Burrell, for example, studied internet users in Ghana. She found that those who lack the financial means to travel turn to the internet to engage 'in fantasies about foreign lands and international travel' (Burrell 2009: 152). In her study of mobile phone use in Mozambique, Julie Archambault (2012) refers to this practice as 'travelling while sitting down' (*viajar sentado*). Digital media have also opened up new interactional spaces for people with disabilities who have hitherto been limited in their physical as well as social mobility. Amanda Baggs, an American activist for neurodiversity, commented in a radio interview in 2006:

> Many of us have a lot of trouble with face to face interaction and are also extremely isolated. Like lots of autistic people, I rarely even leave home. A lot of us have trouble with spoken language, and so a lot of us find it easier to write on the Internet than to talk in person. There is a lot of us we might not be able to meet anywhere else but online, and so that's been a lot of where we've organized. (Ginsburg 2012: 102).

For those who are physically and socially mobile, on the other hand, the technological object itself can become a symbol of their lived mobility. In 2006, I interviewed Thando, a South African woman in her late twenties. Thando grew up in a village in the Eastern Cape, a largely rural part of South Africa. When Thando turned sixteen, she joined her mother, who was working in Cape Town. More than a decade later, now a successful urban professional, she remains in contact with family and friends in the village, texts and calls regularly, and visits every December holiday. Her phone, basic but functional, is not only the object that allows her to be away and to stay in touch, but also symbolizes her status as a young professional and her nostalgia for home. The screen of her phone shows the picture of a traditional amaXhosa home surrounded by pastures, and the ringtone is the morning call of a cockerel, an auditory reminder of village life (Figure 2.1). Here we see how the media object itself, and not only the uses to which it is put, allows for the construction of a particular identity, articulating mobility as well as a desire for home and belonging, her moorings.

An important aspect of work on mobility is its focus on materiality and the objects that make mobility possible. Influenced by the work of Bruno Latour and others, John Urry (2007: 50) emphasizes the complex socio-material relationships that enable and facilitate various mobilities. Thus, different historical and social conditions produce not only different forms of movement, but also different mobility systems, that is, material structures that, in interaction with human actors, support mobility. These include horses, streets and cars, suitcases and shopping bags, as well as pens, paper, computers and mobile phones.

Anthony Elliott and John Urry (2010; Elliott 2013) describe mobile phones, tablets and lightweight laptops as 'miniaturized mobilities'. They not only allow us to communicate across time and space, but are, because of their small size, optimally designed for individual movement and portability. Miniaturized mobilities allow us person-to-person connectivity as well as personalized forms of media engagement while being 'on the move'. This includes entertainment – books, videos and online newspapers – as well as new forms of spiritual engagement and political activism. For example, so-called 'pocket Islam' applications have become emblematic of a modern Muslim lifestyle (Barendregt 2012). These applications include a digital version of the Koran, written as well as recited, a compass that points to Mecca, and prayer times announced with an *Azaan* ringtone (similar applications are available for Christians, Buddhists and Hindus; see Bell 2006 for an overview of techno-spiritual practices). The portability of mobile communication devices has

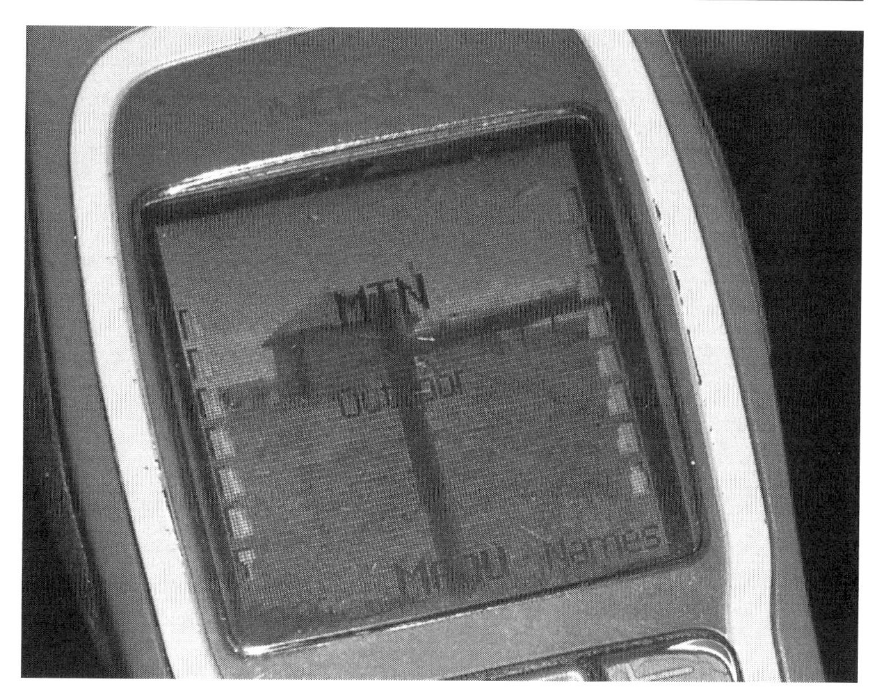

Figure 2.1 Thando's phone: a symbol of mobility and home (Cape Town, South Africa, 2006)

also transformed political activism. Thus, not only do protesters, whether in Turkey in 2013 or during the London riots of 2011, carry mobile phones, but these have become integral to the protest itself. Updates are posted on Twitter as the events unfold, and photos and videos are shared with global audiences online (Schultze and Mason 2012; Rheingold 2002).

The sociologist Tim Dant (2004) has argued that the close integration of technological objects with human activities creates human–object assemblages. He explores this idea with respect to cars and asks whether we can ever write a social and cultural history of cars without considering those who drive them. Who/what pollutes our cities: cars or their drivers? Dant answers this question by proposing a new type of social being, the driver–car, an assemblage of human and object that produces social actions such as driving, parking, speeding, transporting, polluting and so forth. Drivers without cars could not do any of this; nor could cars perform such actions without drivers. The assemblage is necessary, not incidental. A similar line of thinking underpins Donna Haraway's ([1985] 1991) 'Cyborg Manifesto'. Haraway argues that communication technologies are redefining the boundaries of our bodies, and are dissolving the old philosophical dualism of nature vs. technology. While cyborg is a broad cover term that refers to human–machine assemblages in general, Dant's idea of the driver–car encourages us to think more precisely about how specific technologies combine with human agents to produce particular types of actions. Other possible assemblages are the citizen–gun (discussed by Latour 1999), who/that can injure or kill people; the person–remote, who/that can control the program choices on TV; and, in the context of this book, the speaker–mobile, who/that is able to communicate with physically absent others through a variety of mobile media options, voice, text and image.

Typically such assemblages are temporary, continuously disassembled (when I leave the car) and reassembled (when I enter the car). However, in the case of the speaker–mobile we seem to be moving toward a stage of near permanence, of being 'always on', of being in 'perpetual contact' (Baron 2008; Katz and Aakhus 2002). Thus, whether on the metro in Singapore (Figure 2.2) or relaxing at Hamayun's Tomb in New Delhi (Figure 2.3), mobile technologies are always with us and link us to others.

An important concept in these discussions is the phenomenological notion of embodiment, that is, the way in which material objects become part of our bodily memory. Thus, we develop specific ways of entering a car, sitting down in its restricted space and engaging in the process of driving, changing gears, looking out to avoid obstacles and dangers, and processing complex sets of information as we accelerate. Embodiment is also important in our interactions with mobile communication. The ongoing miniaturization of these technologies – the fact that they are getting smaller and lighter and no longer require ongoing connection to wires

Figure 2.2 Singapore (2013): passing time on the metro

Figure 2.3 New Delhi, India (2013): passing time at Hamayun's Tomb

and cables – has facilitated the integration of such devices into the very texture of our lives, making them an extension of our physical bodies. Figures 2.4 and 2.5 illustrate embodiment with two photographs taken in South Africa. In Figure 2.4, the phone is almost invisible. It fits snugly into the hand and is not carried – away from the body – in the spacious bag. Figure 2.5, shows the familiar bodily position of someone reading or typing a message on the phone: the body, shoulders slightly bent, is turned toward the phone, which sits comfortably in the hand. It is not only the materiality of the phone that shapes such bodily practices, but also the social context in which we find ourselves. Being able to carry the phone visibly on the body is not always possible and relies on being in a safe space. If one feels threatened or worried about safety, devices are more likely to be hidden, and the resulting practices are different.

The American psychologist Sherry Turkle, who has studied human–machine interactions since the 1980s, introduced the notion of the 'tethered self' (1995, 2008, 2011) to describe human–machine entanglements and the resulting emotional

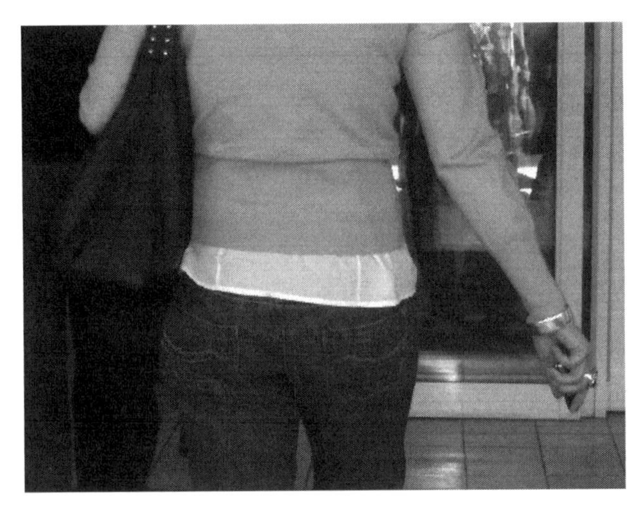

Figure 2.4 Embodiment of mobile phones I (Cape Town, South Africa 2010)

Figure 2.5 Embodiment of mobile phones II (Cape Town, South Africa 2010)

attachments between self and object. She argues that initially, that is, in the 1980s, personal computers were experienced as a 'second self'. They were important to who we are, or would like to be, but were ultimately separate from us (Turkle [1984] 2005; see also Chapter 8). This changed in the twenty-first century. As technologies got smaller and smaller, not only did they become a regular presence in our lives, but also we experienced a new intimacy with machines. Turkle (2007) describes mobile devices – phones, tablets, lightweight laptops – as evocative objects. Like the teddy bear of our childhood, they are objects we live and interact with, and that have the power to evoke complex feelings and emotions in us. Their value is not merely instrumental, allowing us to *do* things; they have become 'companions to our emotional lives' (p. 5). The deep personal relationship many of us have formed with communication technologies is reflected in a joke that made its rounds in 2011. This is the version I was told by a family friend over dinner:

> One day a man finds Aladdin's magic lamp. He starts rubbing it and a Genie emerges. The man, whose marriage is not going so well, asks the Genie to grant him a wish: 'I want my wife to have eyes only for me. I want to be the only one in her life. I want her to sleep always by my side. I want that when she gets up in the morning I'm the first thing she touches and that she takes me everywhere she goes.' And whoosh . . . ! The Genie turned him into a BlackBerry.[3]

The attachments we have formed with technology are most clearly visible when the devices are suddenly not accessible. Many feel a sense of panic when the phone stops working, the internet connection is suddenly gone, or the iPad is acting up. It engenders feelings of disorientation and restlessness, of being cut off, out of touch, lost.

To conclude, digital communication technologies are deeply entangled with the movement of people, ideas and semiotic resources. This section has provided an overview of important aspects of the mobilities debate in the social sciences. This includes a focus on materiality, that is, the objects that enable mobility, and the ways in which some of these objects – phones, tablets and laptops – have become embedded in our everyday lives. The mediated interaction order is no longer an exception, but integral to our lives. And as we engage with each other and with these technologies, we create new types of texts and establish new social practices. How should we approach the creative potential of the speaker–mobile? At this stage I would like to take a step back from digital communication as such and look at creativity and language from a broader and more general perspective. The next section introduces the work of Humboldt and Sapir as a starting point for thinking about creativity, linguistic and otherwise, online and offline.

CREATIVITY AND AGENCY: *TWO CROWS DENIES IT*

As noted in Chapter 1, the theoretical perspective reflected in this book, the 'key' in which it is written, is inspired by a long-standing tradition in the philosophy of language that foregrounds the agency and creativity of speakers rather than the workings of language as a communal, social fact. Language, in other words, is not an object 'out there' that is awaiting description from a linguist. Rather, it is being constantly made and remade by speakers, who are not merely articulating or animating a pre-existing system, a language, but who are languaging, that is, they are 'doing' language (Becker 1995; Makoni and Pennycook 2007; Maybin and Swann 2007; Jørgensen 2008; Blommaert and Rampton 2011). These ideas shape current thinking in sociolinguistics, but also have a long history. They go back to the beginnings of the discipline in the nineteenth century, and have surfaced repeatedly throughout the twentieth century. Yet until recently, they were largely outside of mainstream linguistics and those branches of sociolinguistics that have foregrounded inquiries into linguistic structure.

The language-philosophical writings by Humboldt and Sapir provide an important counterpoint to a preoccupation with order and structure, and direct our gaze instead toward experiment and novelty. Both articulate important sociolinguistic ideas *avant la lettre*, that is, before sociolinguistics emerged as a fairly well-defined area of study. Humboldt and Sapir were linguists-cum-anthropologists, and as such interested in understanding social life, not simply linguistic form.

Wilhelm von Humboldt's language philosophy is based on a fundamental distinction between *ergon* (language as a product or an object) and *enérgeia* (language as activity, language as formation). The distinction is similar to the one made later by Ferdinand de Saussure ([1916] 2013) between *langue* (language as a system) and *parole* (speech). However, while Saussure privileges *langue* as the main focus of linguistic analysis, Humboldt privileges *energeia*, language-in-use or languaging. Humboldt states this clearly in the introduction to his work on Kawi, an ancient language of Java (today usually referred to as Old Javanese), which was published in 1836.

> Language, considered in its true nature, is something which is constantly and in every moment in transition . . . *Language is not a product (ergon) but an activity (energeia).* ([1836] 1998: 174, my translation and emphasis)

Consequently, the nature of language cannot be understood by describing language as a bounded, self-contained system of conventionalized signs, an approach that defined nineteenth-century historical linguistics and later Saussurian structuralism. Humboldt was unambiguous in his dismissal of such work. For him the resulting grammars and dictionaries were nothing but a 'monstrosity of scientific dissection' (*Machwerk wissenschaftlicher Zergliederung*), leading to the destruction (*Zerschlagung*) of language, not its description. Rather, language should be seen as part of our lived experience, inherently social and dialogic, always oriented toward a real or imagined other. He developed this idea in *Über den Dualis* ('About the Dualis'), where he argues that dialogue and sociability (*Geselligkeit*) are part and parcel of the very nature of language.

> All speaking is based on dialogue . . . the word, born in solitude, too strongly resembles a mere imaginary object, language too cannot be realized by an individual, language can only be realized socially. ([1827] 1994: 164–5 my translation)

Meaning, in other words, is always and necessarily emergent, the creative result of the semiotic work of those participating in interaction. This early dialogic view of language and meaning (anticipating the work of Bakhtin; see Chapter 6) stands in stark contrast to the *sender → message → receiver* model – also called the conduit model – which depicts communication as a straightforward process of encoding and decoding meaning. The conduit model, which is famously depicted in Saussure's talking-heads sketch, was dominant for most of the twentieth century, in academic linguistics as well in folk understandings of how language works (Harris 1981).

Speaker agency is central to Humboldt's language philosophy. Yet such agency is never boundless and language is both freedom and limitation. I can speak/write only because others have spoken/written before me, because there exist conventions of usage. But, at the same time, my speaking is always more than a reproduction of these conventions, and in the act of speaking/writing I appropriate, adapt and change language. Humboldt emphasizes the fundamental tension between the energy and force of the individual (*Kraft . . . Gewalt des Menschen*) and the power of language (*Macht der Sprache*) in his work on Kawi as follows:

> Language belongs to me because I produce it in my very own way; and since the basis of this lies in the speaking and having-spoken of all previous generations . . . it is the language as such that limits me . . . the study of language must recognize and honour the phenomenon of freedom, but at the same time trace carefully its limits. ([1836] 1998: 190f., my translation)[4]

Aspects of Humboldt's language philosophy resurfaced in the twentieth century, not only in Germany, but also in the United States, where Franz Uri Boas (1858–1942), originally from Germany and well versed in the philological tradition, established the field of linguistic anthropology. An important link between Humboldt's work and early linguistic anthropology is what became known as the (Humboldt–)Sapir–Whorf hypothesis; the idea that human cognition and perception are embedded in structures of specific languages, leading to a multiplicity of ways of seeing, conceptualizing and assimilating the world around us (Humboldt called this *Weltansichten*, 'worldviews', in his writing; see Deumert 2013).

However, Humboldt's other views on the nature of language were not ignored, and roughly a hundred years after Humboldt described language as the creative activity of individuals, Edward Sapir – a student of Boas – articulated a program for linguistic study that was equally critical of those who approached languages as 'objects out there' awaiting study by linguists. Less dismissive of grammars and dictionaries than Humboldt, Sapir was a prolific field linguist, heavily engaged in linguistic description. However, he too remained skeptical as to the ontological status of the produced grammars. He did not see them as full or exhaustive descriptions of 'a language', but simply as an abstraction, a 'convenient summary' of historically and socially *situated* human actions and interactions (1917: 446). And like Humboldt, he emphasized that the individual – as speaker and listener – is not simply a carrier and animator of language, but has transformative and expressive power. In a programmatic paper, titled 'Speech as a Personality Trait' (1927), Sapir emphasizes the ability of speakers to use language in personal and unique ways. Although linguistic practices are shaped by social norms, speakers are always able to transform and subvert. Sapir writes:

> [S]ociety has its patterns, its set ways of doing things . . . while the individual has his method of handling those patterns of society, giving them just enough twist *to make them his and no one else's.* (Sapir 1927: 894, my emphasis)

Sapir's concern with human agency is linked to his strong interest in the poetic and creative aspects of language. Being an accomplished poet himself, Sapir knew that although a writer cannot have absolute freedom, it is nevertheless true that '[t]he possibilities for individual expression are infinite' (1921: 221).

By foregrounding the tension between that which exists already and the new that is possible, Sapir and Humboldt describe the signifying individual in a way that anticipates the poststructuralist understandings of the subject. In poststructuralist thinking, the individual is not only an actor or agent (as in 'the subject of the sentence'), but also someone who is placed under authority and control (as in 'the king's subjects'). Thus, speakers, and speaker–mobiles, are subjected to pre-existing structures – that which is there, including local and global indexical orders that assign value to certain forms of speech and disvalue others – but also always able to resist, change and subvert these structures (that is, to create novelty; see Kramsch 2009 as well as the discussion of Derrida and Butler in Chapters 5 and 6).

Sapir's ([1938] 1968) most explicit discussion of language and creativity is found in his essay on Two Crows, an Omaha Native American who features in an ethnographic study published by John Owen Dorsey in 1884.[5] Two Crows is described as 'a perfectly good and authoritative Indian' (p. 570), that is, as someone who can be considered knowledgeable about community practices. However, Two Crows is also a bit of a cultural rebel and brusquely dismisses any generalization Dorsey makes about Omaha society and cultural practices. The phrase 'Two Crows denies it' is a frustrating refrain in Dorsey's book, potentially invalidating any observation he has noted down and described as 'typical' of Omaha society.

Sapir uses the character of Two Crows to reflect on the importance of hearing dissent, heterodoxy and contrariness in our data, of listening to the voice of the individual. He notes that Two Crows 'has a special kind of rightness . . . partly factual, partly personal' (p. 574). That is, whatever Two Crows claims and does, even if different from those around him, is obviously true for himself. Sapir then constructs a thought experiment around Two Crows in order to illustrate the potential for individual creativity as well as its ability to bring about social, cultural and linguistic change. What if Two Crows were to exchange the first and the last letter of the alphabet? What if he were to insist that the alphabet starts with Z and ends in A? To most observers such behavior would seem exceedingly strange and Two Crows would probably be considered weird, strange, crazy or maybe, in a more positive spirit, subversive and transgressive. However, if others were to agree with him then his initial act of rebellion might lead to a new alphabetic tradition. Sapir concludes:

'his divergence from custom had, from the very beginning, the essential *possibility* of culturalized behavior' (pp. 571–2, my emphasis).

When I began working on digital communication in South Africa I encountered many Two Crows. As soon as one person suggested a broad generalization about common practices, someone else would play Two Crows and deny that this was indeed so. Ilona Gershon (2010) noted something similar in her work on new media in the United States. The people she interviewed disagreed frequently on the sociolinguistic conventions that guide the use of digital media. One reason for such disagreements might be the relative newness of these technologies and the speed of change; practices just haven't yet had time to settle down. Yet if we take the reflections by Humboldt and Sapir seriously, then dissent and disagreement are not necessarily a passing phase, but integral to the workings of social life. Dissent – which is often experienced as disruptive to the social order by those in positions of authority (including academics trying to write a monograph) – is also the basis for creativity. By doing otherwise than expected, by challenging society's 'patterns, its set way of doing things', new forms are created. Following Sapir's reflections on Two Crows, any behavior – however unusual or mercurial – has the *potential* to become a new convention. It is important to emphasize that unpredictability and openness are not limited to digital communication, and are a feature of language and language change more generally. They have been discussed, for example, in the field of language contact studies (e.g. Thomason 2000) and in work on secret languages and other forms of linguistic manipulations (Storch 2011), and have been emphasized in research on the creative and artful sociolinguistic practices of, especially, young people (e.g. Alim 2011).

Having charted the broad theoretical terrain – virtuality, the archive, mobility and creativity – the next section considers methodology and ethics: how should we study digital practices?

METHODOLOGICAL AND ETHICAL CONSIDERATIONS

Ethnographic approaches have been important in the study of digital interaction and come in two main versions: (1) studies that focus on an offline fieldsite, and (2) studies that focus on an online fieldsite.

1. Researchers who base their work on an offline fieldsite select a particular group that conforms to the traditional anthropological requirement of physical and social propinquity, such as a neighborhood, a school or even a nation. They then study how members of this group use communication technologies in their everyday lives, how offline and online practices intersect and complement one another. Good examples of this are Heather Horst and Daniel Miller's (2006) ethnography of mobile phone communication in Jamaica, and Daniel Miller's (2011) study of Facebook in Trinidad. Researchers know the offline identities of participants, and are able to work in a fairly traditional ethnographic manner.

2. Researchers who focus on an online fieldsite take an environment such as Facebook or the World of Warcraft and study those who interact with one another on this platform. The focus is firmly on the virtual, the practices it supports and the interactions that occur. An example of this approach is Tom Boellstorff's (2008) ethnography of the virtual reality site Second Life. Researchers mostly interact with participants' avatars and know their offline identities only if participants are willing to reveal them.

Both approaches are classically ethnographic in that they focus on a particular *place*, whether concrete or virtual, and use well-established field methods (such as participant observation and interviews). However, Guobin Yang (2003) has argued that strongly localized approaches are ill-equipped for exploring the essential openness, diversity and connectivity of digital media. He suggests a more exploratory approach, which he calls 'guerilla ethnography' and describes as follows:

> Get involved [in particular sites and with particular technological devices] but be ready to move around in the networks. Explore links. Take abundant notes, download information, and when appropriate, post questions and solicit answers. Sit back and think about the larger picture. Return to selected sites for deeper exploration. (p. 471)

This approach gives due recognition to the networked nature of digital communication as well as the fact that different devices and platforms might support and encourage different practices. It also gives new meaning to the old notion of 'armchair linguistics'. It is possible to discover a lot about the digital world by just keeping one's eyes on the screen.

Yang's guerilla ethnography can be illustrated by imagining a study that wanted to investigate expressions of death, grief and loss on the Internet. One could, for example, start by looking at Facebook groups that host remembrance pages, such as the 'Memorial Wall for friends who left too soon', as well as pages that commemorate specific individuals. Following this, one might explore further sites, such as blogs written by family members and friends, or webpages that provide counseling and advice on how to cope with death, or search Twitter for postings. And then there are sites such as virtualeternity.com. The site allows users to build avatars of themselves that will live on – virtually on the website – after the creator's physical death. Although the scope of the study is potentially vast and rhizome-like in shape, a broad field of potentially relevant sites can be charted quite quickly. In addition to so-called log data – screen shots of webpages, collections of Facebook updates and comments, chat conversations and tweets – researchers would typically carry out interviews with those who contributed content, or ask them to complete media diaries, where writers reflect on their digital practices (Androutsopoulos 2008; Barton and Lee 2013: 164ff.). This allows emic interpretations to surface, that is, the meanings speakers assign to their own practices and the practices of others (as opposed to etic interpretations, i.e. those of the analyst).

Yang's approach to online ethnography is similar to what emerged under the name of multi-sited ethnography in the 1990s (Marcus 1995; Gatson 2011). Multi-sited ethnography, which is closely connected to the mobilities paradigm discussed above, moves away from a focus on single sites, and foregrounds the connections of people, meanings and objects across different spaces. An approach that is similar to multi-sited ethnography but slightly different in emphasis is what Monica Büscher and her colleagues (2011) call 'mobile methods'. Here the focus is not so much on multiple sites and their connections; instead, the researcher moves along with the object under investigation. Julia Pfaff (2010), for example, followed a mobile phone from Zanzibar to Dar es Salaam. She showed how it passed from owner to owner and was put to different communicative uses in different contexts (follow-the-thing approach). Another possibility would be to follow a particular text or image, to see how it moves across the internet, is reposted and modified (follow-the-text approach; see Chapter 5), or to follow a particular individual and document how she engages with digital media at different times and in different places (follow-the-people approach).

The internet is not only an object of study, a huge corpus of language-in-use, but also a source of information about itself. For example, the user-generated Urban Dictionary offers emic explanations of popular online expressions, YouTube provides us with viewer statistics for each video, the Alexa rankings (alexa.com) track the popularity of individual websites, and Opera (opera.com) publishes regular reports on the mobile-centric internet. For the United States, the Pew Research Center's Internet and American Life Project gives researchers free access to a range of statistics and scientific reports (http://www.pewinternet.org), and on a global level, the International Telecommunications Union (www.itu.com) offers a vast statistical archive. With regard to log data, Facebook's API (application programming interface) allows one to download public Facebook status updates and comments (similar applications are available for Twitter). A useful tool for longitudinal web research is the Way Back Machine, which allows one to call up earlier versions of a webpage (http://archive.org/web/web.php). This is illustrated in Figure 2.6. The first image shows the retrieved 1997 webpage of the University of Cape Town, and the second image the current version. The example shows clearly that web design has come of age. It has moved from mainly text designs to more complex, multimodal design, that is, a greater use of images and embedded videos.

And finally, it is essential to consider ethical challenges. If we accept that the actual and the virtual are deeply entangled, then the basic ethical principles for offline research should also apply to online research: the right to informed consent, anonymity and the protection of privacy. This is particularly important when dealing with data from restricted-access networks or private messages. Many researchers anonymize not only people's names but also their nicks (or handles), since the latter index particular virtual identities, just as personal names point to specific offline identities (see Chapter 8). In addition, all identifying information should be removed. Thus, it would not really be sufficient to anonymize data from my Facebook wall by stating 'written by a Cape Town linguist in her forties'. The

Best viewed using Netscape v.3+ or MS Internet Explorer v.3 on a monitor displaying 256 colours.

Welcome to the
UNIVERSITY OF CAPE TOWN
Promoting Excellence with Equity

Founded in 1829 as the South African College
and constituted by Charter in 1918 as the University of Cape Town.

News and Announcements
EXAM RESULTS: **Updated regularly**

Search for staff e-mail addresses
Vacant Posts

- Introducing UCT
- UCT's Mission Statement
- The University Transformation Forum (UTF)
- Studying at UCT
- Information for International Students
- Academic: Faculties ‖ Departments ‖ Research
- Conferences ‖ Inaugural Lectures ‖ Public Courses
- The UCT Libraries
- Information Technology Services
- Regional Collaboration

- Managing and Administering UCT
- new Policies and Procedures
- PRISM
- Development and Public Affairs
  - *Monday Paper* -- updated weekly
  - Alumni/ae Affairs
- Theatre, Music and Art -- Updated regularly
- The UCT Club
- UCT Press
- UCT Student Organisations
  - SRC Elections 1997

OTHER LINKS

- Learned / Professional Societies and Associations
- The City of Cape Town ‖ The Western Cape
- Education Resources ‖ UCT Careers Office
- Internet Services ‖ WWW Starting Points
- Awards to UCT Web Site

All mail should be addressed to:
The Registrar
University of Cape Town
University Private Bag
Rondebosch 7701, South Africa

General inquiries:
Telephone: (+27 21) 650 9111
Facsimile: (+27 21) 650 2138 / 4040

Individual departments and sections have their own telephone and fax numbers,
which are listed where appropriate.

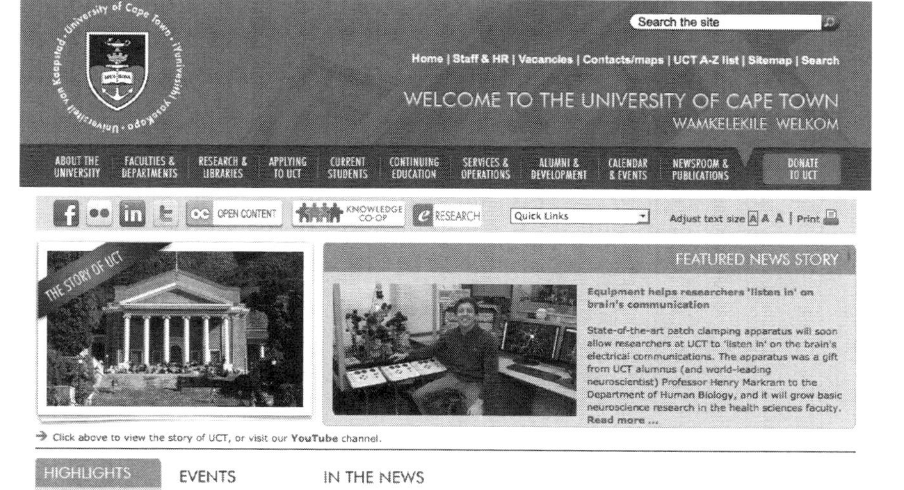

Figure 2.6 The webpage of the University of Cape Town: 1997 and 2013 (www.uct.ac.za; the 1997 version was recovered using the Way Back Machine)

community of linguists in South Africa is small, and this information would be sufficient for many people to identify the person as me (D'Arcy and Young 2012; Schultze and Mason 2012).

The situation is somewhat less clear-cut on sites that are publicly available. Do we need to obtain informed consent and anonymize our data if we want to analyze Twitter or comments posted on YouTube videos? And what about people who have not activated any privacy settings on their Facebook page? In the past researchers have often equated technical accessibility with publicness, and used such data quite freely and extensively. Yet people do not always consider their data public and can use such spaces to engage in fairly private conversations. It is here that researchers need to tread carefully. Privacy needs to be understood as contextual and emergent, and we need to evaluate each case on its own merits. Thus, while some Twitter posts do not carry any identifying or personal information, and are similar to public blogs, other tweets might be quite personal and not suitable for citing without consent. Anonymity is another issue. Generally, anonymizing one's data is advised, even if the data is considered public. While this is general practice, however, it does not actually succeed in protecting privacy and guaranteeing anonymity. If the text is quoted verbatim, then it is usually possible to locate the original context (and the author) via a simple web search.[6] This is a quandary that cannot be resolved easily (see also Chapter 4).

Moreover, there are some contexts where authorship needs to be acknowledged. Obvious examples of this are professional blogs or Twitter accounts by well-known public personae. If in doubt how to best protect and respect those who are at the center of our research, we should apply what one might call the 'golden rule' of research ethics thoughtfully and with sensitivity: how would I feel if someone treated me in the way I am proposing to treat others?

CONCLUSION: THE MEDIATED INTERACTION ORDER

The ubiquity of new media in today's world encourages us to reconsider Goffman's comment that mediated interactions are nothing but 'reduced versions of the primordial real thing'. Although mediated interactions are no less real than body-to-body interactions and can be experienced as deeply personal (as illustrated by the 'rape in cyberspace' example), they are also different from body-to-body interactions, and it is important that we consider these differences carefully.

Mediated interactions are virtual and transcend the time-space constraints of co-presence. Virtuality allows us physical mobility, which is further aided by the portability of communication technologies and their close integration into our daily lives. The fact that texts can be archived and retrieved facilitates their mobility. It is thus possible to repost and remix existing texts, images and videos and make them work in new contexts, both local and global. Although Goffman saw mediation as marginal to the interaction order, he was nevertheless interested in mass-media communication (radio and advertising; Goffman 1981 and 1979 respectively). In these writings he observed that mediated talk intensifies aspects of body-to-body interaction. By allowing for editing, it creates the possibility of taking out the 'dull footage' and emphasizing 'colourful poses' (1979: 84). In other words, mediated interaction, freed from the constraints of real-time processing, allows time and space to revise and edit, and thus supports creative and artful sociolinguistic practices. I will return to this in more detail in Chapters 5–8.

An often-mentioned challenge for the sociolinguistic study of digital communication is not only the scale of such practices, ranging from local to global, but also the speed of change, that is, the fact that the medium and the way in which it is used are constantly being reconfigured. Change is not simply about new technologies that allow us new ways of interacting, but also about how their material footprint transforms the social order, the structures of everyday life and the diverse *-scapes* of our world (landscapes, ethnoscapes, mediascapes, technoscapes, soundscapes, etc.; Appadurai 1996). Consider, for example, the history of so-called container phones in South Africa. A container phone was a refurbished shipping container that was fitted with public phones and run as a franchise business. In these phone shops, customers could make phone calls for less than a third of the commercial rate for mobile phones. Although using one's own phone was seen as preferable – since it allowed privacy and mobility – limited finances meant that many people would use a stationary public container phone to *make* calls, but would continue to *receive* calls on their mobile phone. In 2009, container phones were abundant in socio-economically marginalized neighborhoods in South Africa. Now they are disappearing. As many people are becoming more affluent, they are willing to pay just a bit more for privacy, mobility and convenience. However, the containers have not disappeared from the urban landscape. Some are vacant and closed, but many have been transformed into hair salons, beauty parlours or even restaurants.

Figure 2.7 From public phone container to beauty parlor (Mthatha, Eastern Cape, South Africa, 2012)

In Figure 2.7, the original design of the local Cell C network provider, depicting people communicating, still adorns the container, which now houses a pedicure studio.

The next chapter focuses more closely on questions of materiality. What kinds of actions do digital media facilitate, and who actually has access to them?

NOTES

1. The word 'avatar' comes from Sanskrit (*avatāra*) and originally described the descent of a deity – a virtual being – to earth in an incarnate, physical form. In virtual worlds the semantic movement goes in the opposite direction, from physical body to virtual representation.
2. In October 2013, Snapchat Stories was launched. It allows one to post series of photos, a mini-narrative, which will self-destruct not after 10 seconds but after 24 hours. Thus, the here and now is stretched to here and tomorrow, but the basic principle – no archiving – remains.
3. The joke exists in two versions, with either the husband or wife requesting a wish and being turned into a BlackBerry.
4. Chomsky (1965) appropriated Humboldt's idea of language as *energeia* and saw it as an equivalent to his mentalist idea of (generative) competence, an interpretation that has been rejected by Humboldt scholars. Chomsky's notion of creativity is fundamentally opposed to Humboldt's thinking, since Chomsky recognizes only *rule-following* creativity, not *rule-changing* creativity (see also Chapter 9).
5. The Omaha are Native Americans who now live on a reservation in northern Nebraska.
6. The most recent ethics guidelines of the Association of Internet Researchers (AOIR) can be found at aoirethics.ijire.net.

Chapter 3

Affordances and access

> Hey #civicmedia, we need to talk seriously about digital access inequality as an ongoing, structural impediment to truly democratic commons.
>
> Twitter 2013

INTRODUCTION: THE POLITICAL ECONOMY OF DIGITAL COMMUNICATION

Everyone knows the saying 'talk is cheap', meaning that it is easier to say something than to do it. To talk to someone body-to-body is not only cheap, it is better than cheap: it is free. The same is not true for digital communication. Online practices such as chatting, blogging or texting cost money and are thus deeply embedded in the larger political economy.[1] For example, many platforms that have become emblematic of Web 2.0 are profit-making ventures. North America has so far dominated the market, although hardware and communication applications designed in Asia are gaining ground (e.g. LINE, WeChat). Not only does participation occur within commercially designed architectures and require investment in hardware and data packages, but that which is created becomes economically productive for those who provide the infrastructure. On Second Life, a site studied by the anthropologist Tom Boellstorf (2008), users create the very world they inhabit by building virtual houses and landscapes; at the same time they pay membership fees to Linden Lab for residing and interacting in the world they have fashioned. YouTube, Twitter and Facebook might not be charging membership fees, but they too are business ventures: each new video on YouTube, each status update on Facebook, each posting on Twitter makes those sites more attractive to advertisers. (The political economy of digital media is discussed in detail by Fuchs 2014a and 2014b.)

In this chapter I explore the political economy of mobile communication by focusing on the global distribution of technological objects. I will consider who has access to what types of technology, and look at the ways in which technology has been creatively appropriated, manipulated and domesticated in different societies and by different groups of people. This is commonly discussed under the heading of the digital divide. In a nutshell, digital access and use – like so many other things – are unequally distributed: some people have ample access to the internet and digital media (such as phones, tablets and laptops; they are commonly called the haves); many have some access (the have-less people); and a shrinking proportion of the world's population has none (the have-nots; Ragnedda and Muschert 2013). Classic sociological variables such as gender, age, social class, income, disability and ethnicity/race structure access and usage within countries as well as globally. However, before exploring questions of global access and use, I will revisit the speaker–mobile (discussed in the previous chapter) as a particular type of social actor and consider the types of communicative actions that digital technologies, as artifacts with particular design features, allow us to perform.

COMMUNICATIVE AFFORDANCES

Scholars of new media typically begin their discussion of the social consequences of technology by contrasting two approaches: determinism and constructivism. A strictly deterministic perspective argues that new technologies 'cause' social change: their mere presence transforms social institutions (the macro-level of social life), as well as relationships between individuals and our sense of being-in-the-world (the micro-level of social life). Determinist perspectives have a long history. They go back to antiquity, when Plato expressed concern about the negative effects of writing and reading on memory, creating forgetfulness and a reliance on secondary (read-about) experiences (Harris 2000: 19ff.). In the twentieth century, Marshall McLuhan (1964) expressed a similar view when he argued that the messages we communicate are shaped and even altered by the technologies we employ. He coined the much-cited phrase 'the medium is the message'. Determinism is also visible in ongoing popular debates about whether computer games are 'causing' violent behavior in young people, or whether texting is 'destroying language' (see also Chapter 7). Such arguments are often framed as before-and-after narratives, and technological objects are positioned as active agents: 'Google *makes* us stupid' or 'robots *put* people out of work'.

Social constructivists, on the other hand, emphasize the power of human agency and maintain that technological objects should be thought of as akin to texts that are interpreted and made useful in the specific contexts of our everyday lives. This perspective is associated with the work of the British cultural scholar Raymond Williams (e.g. 1974). Williams argues that the effects or uses of a technology cannot be predicted, that people will always bend the technology to suit their own needs and desires, and that technological objects present us with various possibilities for action. Such possibilities for action are referred to as affordances, a concept introduced by James J. Gibson when writing about the psychology of perception (1979; discussed in Hutchby 2001: 26ff.). The idea of affordances is not limited to technological objects, and all objects, the material world in its entirety, allow human beings to *perceive* possibilities for action. Consider something as mundane as a rock. If the rock is large enough, then I can sit on it and rest; if it is portable, then I can display it as part of my rock collection; I can even throw it (for fun, for exercise or with destructive intent). It is the very materiality of the rock, its shape, size and weight, that enables me to perform certain social actions, but not others. I cannot eat it, I cannot mould it, I cannot cut it etc. However, the mere presence of a rock does not, in itself, bring about or 'cause' certain types of human behavior. I might also ignore the rock and walk on.

Another example is the music capacities of mobile phones. Most phones have integrated mp3 players, and when one purchases a phone it comes with a set of earphones, which allow for private listening, at home as well as in public spaces. Yet in South Africa, one often sees groups of teenagers sitting together, listening to music, even dancing to it, with the volume of the mobile phone turned to maximum capacity. Something similar has been described for China, where farmers, working the fields in groups, enjoy listening to music or the radio together. They are increasingly using their mobile phones for this, again with the volume turned up as high as possible (Oreglia and Kaye 2012). In both cases users have turned an individualized technology into a social machine, similar to a boom-box, just more portable and always ready to hand. This use of mobile phones was, however, not anticipated by developers. The in-built speakers are rarely loud enough for social listening, and the experience, while better than nothing, is also frustrating. This means that although technologies don't dictate how we should use them, their design puts limitations on our creativity. Mobile phones can become mini-boom-boxes for small groups of friends or co-workers; however, they are unlikely to be a satisfying solution for a large party (unless connected to an external speaker).

Phones and laptops not only afford us mobility (as discussed in the previous chapter), but also carry with them interactional affordances, that is, particular ways of interacting with others. Importantly, they allow us choices of *how* we want to be in touch with others. Do we want to call or send a text message? High-end smartphones increase the range of options even further: voice call, video call, text, tweet, Facebook message etc. Whereas body-to-body communication as well as voice or video calls require real-time interactivity, text-based interactions allow us to desynchronize interactions. Even in contexts where there is an expectation of an immediate reply, we can usually delay our response. That is, we can step out of the interaction order by offering explanations such as 'I was in a meeting', 'no network' or 'battery was flat'. Naomi Baron (2008) has argued that the affordances of digital technology allow for 'volume control'. If we want to retreat from an interaction we can do so more easily than in body-to-body contexts. This ability to delay, to opt out of real-time interactivity, affords us the opportunity to think carefully before answering a pressing question, diminishes the face-threatening potential of a rejection, and allows us to control the presentation of the self more tightly than is possible in the ongoing flux of communication (this will be discussed further in Chapter 6). The lack of co-presence also affords the possibility of anonymity, of presenting ourselves as different from the person we are in actual life and thus constructing an imaginary identity (this is discussed further in Chapter 8).

However, volume control is not absolute: there are limits to how long we can delay answering a call or a text for without raising suspicions. In close relationships, especially, temporalities of response need to be carefully managed. Example (1) is a text message sent by Euphrasie, a Côte d'Ivoirian student in her early twenties, to her boyfriend, who has not been answering her calls for almost a day. Starting to get suspicious, she texts him. (The message is first given in the original orthography, then in standard French, followed by a translation into English.)

(1) Mè prkw 2p8 18h tn phone è fermé. Jèspè k tè pa en trin 2 fè dè choz bizzar
'Mais pourquoi depuis 18h ton téléphone est fermé ? J'espère que tu n'es pas en train de faire des choses bizarres'
'But why was your phone closed for eighteen hours? I hope you are not doing bizarre things' (Côte d'Ivoire, 2011)

A switched-off phone, an unanswered message or call can become incriminating evidence in intimate relationships, and digital technology also enables new forms of surveillance. Lovers check call logs and saved messages; they might scrutinize Facebook comments and lists of friends. Yet, at the same time, new media can also provide freedom from surveillance, and offer a space for unmonitored interactions, especially in contexts where such interactions are not possible body-to-body. Thus, in societies where courtship between young adults is heavily scripted and socially policed, mobile phones offer an opportunity to engage discreetly in private and intimate exchanges (see, for example, Lamoureaux 2011, for Sudan).

It is not only the temporalities of response that carry meaning, but also the choices made between different media. Ilana Gershon (2010) argues that people hold media ideologies, that is, they have beliefs about technologies and the meanings that are implied in the choice of a particular technology. For example, can you break up a relationship via text? Should you call? Is email an option? What about Skype? Or must it be body-to-body? Many people consider break-up by texting the worst possible choice. Texts are simply too short to allow for proper explanations, and are often considered a medium for playful intimacy, not for heartbreak and pain.[2] So maybe email, which allows one to explain one's reasons carefully – similar to an old-fashioned 'Dear John' or 'Dear Jane' letter – is OK? When I suggested this to my South African students in 2012, they responded much like Gershon's American students, with laughter. For them, in their early twenties, email is too cold, too businesslike. It is a medium solely for professional communication, that is, to contact lecturers, tutors or employers, not for friendship and romance. Others, however, might not share this view. For me and many of my generation, email is professional as well as personal. That is, I can send formal letters via email, but I can also send also short notes to friends or engage in romantic interactions. My own media ideologies, and consequently uses, thus differ from those of my students. There are also differences in the way applications are used. For some, Twitter is a tool for micro-blogging, allowing one to publish one's thoughts to a potentially global audience. For others, Twitter is about interaction. Through tweeting, retweeting, hashtagging and responding to people, they create community and a sense of belonging (Zappavigna 2014; see also Chapter 8).

In other words, the design of the platform and its technological affordances do not determine practices. Let us look at another example. An important linguistic feature of many body-to-body interactions is back-channel responses (*hm hm, yeah, OK*) and overlapping speech. These are forms of listener behavior associated with conversational support. Back-channeling and overlapping speech – especially full or partial reformulations of what was said – are also important in counseling, where they are known as active listening. In a study of online counseling sessions Susan Danby and her colleagues (2009) found that such active listening – which has been well documented for body-to-body as well as phone counseling – did not occur in the chat protocols they analyzed. Although the sessions used a chat program that allowed for synchronous, real-time interaction, the resulting conversations were more like monologues, with little evidence of the therapeutically important practice of active listening. Why did this happen? In chat interactions, unlike in spoken interactions, it is not usually possible to see how the turn produced by one writer unfolds as it is being crafted. That is, we do not see the words as they are being typed, and while one person is typing, the other person will merely see a line at the bottom of the screen, 'X is typing'. And although it would be possible to type *hm hm* or *OK* at regular intervals, these utterances would not respond to anything, as the story that is being told is still invisible to the listener. In one of the extracts discussed by Danby and her colleagues, the client/patient describes her current situation in two extended turns, lasting almost 10 minutes (time elapsed while typing). Only when the client/patient signals explicitly that she is now finished ('sorry I think that is all') does the counsellor respond. This is quite unlike what we know from body-to-body interaction, which tends to be more dialogic in structure.

However, as argued above, it is important not to think deterministically. A much more dialogic style of interaction is illustrated in example (2). The chat extract shows eight conversational turns in ten minutes. Although Laura (early forties, professional) dominates the conversation, she delivers the update about her plans to buy a new laptop in short chunks, frequently pressing the enter key (indicated by ↵) Tony (her partner, also mid-forties, professional) gives supportive and affective feedback, using short turns as well.

(2)	4:59	*Laura*	just quickly checking the computers ↵
			i think I found one – just read the review ↵
			it sounds good and has a good price, just under 8000 ↵
	5:01	*Tony*	good, what did you find? ↵
	5:02	*Laura*	it is n HP pro book 4730s ↵
			the review sounded very good ↵
	5:05	*Tony*	cool <3 ↵
	5:06	*Laura*	will order it tomorrow, i need to get the fund number <3 ↵
	5:07	*Tony*	ok, ↵
	5:09	*Laura*	shall we work till six and then leave? <3 ↵
	5:09	*Tony*	ok ↵
			<3 ↵

(Gmail chat, Cape Town, South Africa, 2012)

Thus, chat turns can be kept short (by pressing the enter key frequently). This not only heightens the sense of immediacy, but also allows the other person to comment on the narrative as it unfolds, and the technology itself is neither monologic or dialogic.

Technologies, however, can only afford us possibilities for action if we have access to them, and such access is unequally distributed across the world. The next section looks at the shape of the digital divide. I identify areas in the world where digital access is plentiful (the haves), areas where there is some but limited access (the have-less people) and those areas where access is very restricted and many remain unconnected (the have-nots).

DIGITAL INEQUALITIES: GLOBAL ACCESS AND USE

In the late 1990s and early 2000s, proponents of the so-called ICT4D (Information and Communication Technologies for Development) paradigm rallied around the utopian and essentially deterministic idea that digital connectivity by itself can lead to socio-economic growth. The basic idea behind such thinking is that access to technologies, such as the internet, facilitates the circulation of information. This information can then be turned into entrepreneurial opportunity, the improvement of life chances (such as better health or educational advancement) and democratic participation more generally. However, many ICT4D projects failed dismally. An example of this is the Gyandoot[3] project, which was set up in 2000 by the Indian government to provide digital access to villagers in Madhya Pradesh (India). Over thirty tele-kiosks delivered, via a government-sponsored intranet service, information on agriculture and markets, health, education, women's issues, poverty alleviation and government services. Yet villagers showed little interest, and a study of eighteen kiosks over a period of two years found that, on average, they served a mere 0.62 visitors a day (Mazzerella 2010). Lack of interest might have been a result of what was on offer at the kiosks. Like many other projects, Gyandoot restricted digital access to what government officials considered 'useful' and 'beneficial'. For example, villagers couldn't access news sites, play games, watch videos or chat with people far and near. About a decade later a different approach was taken in two village-based projects in southern Africa. In Macha (Zambia) and Dwesa (South Africa) wireless internet access was made accessible through community terminals. Villagers were not limited as to the sites they could access and had the entire internet at their disposal. When David Johnson and his colleagues (2011) carried out a traffic analysis in the two villages they found that residents had become active users of the internet, but not necessarily of 'useful' services. Most web traffic came from social network(ing) sites, especially Facebook.[4]

The limited success of many ICT4D projects reminds us that individuals and local communities will always establish their own digital cultures, focus on aspects of the technologies that matter to them, and use them in ways they find desirable. Thus, digital technologies are not only useful objects, which allow connectivity and access to information. They are also expressive objects, and by owning and using them in certain ways we constitute ourselves as certain kinds of people. The cultural images associated with telecommunications and new media are far removed from the utilities of government forms, title deeds and educational material. Access to communication technologies indexes success, modernity, affluence and global connectivity. Mobile technologies have emerged as consumer items and status symbols, a 'new vehicle of identity and identification for all walks of life' (De Bruijn et al. 2009: 14; see also Figure 3.1). These ideologies shape usage as well as sociolinguistic practices.

Yet many people in the world lack the means to purchase the technology they desire. In order to obtain a global summary of digital inequality, access as well as usage, it is useful to consider groups of similar countries together.

Binary distinctions – between modern and traditional, West and non-West, developed and underdeveloped, core and periphery, First World and Third World – have a long tradition in the way social scientists think about the world. Since the

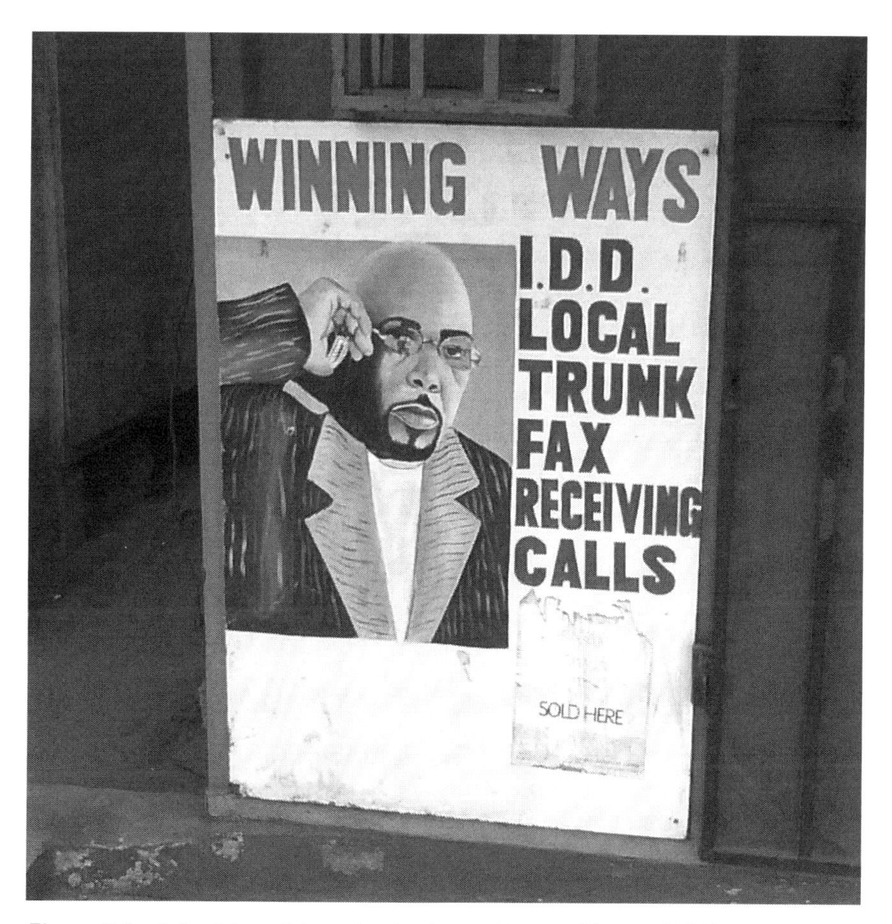

Figure 3.1 Advert for a telecenter business (Accra, Ghana, 2009)

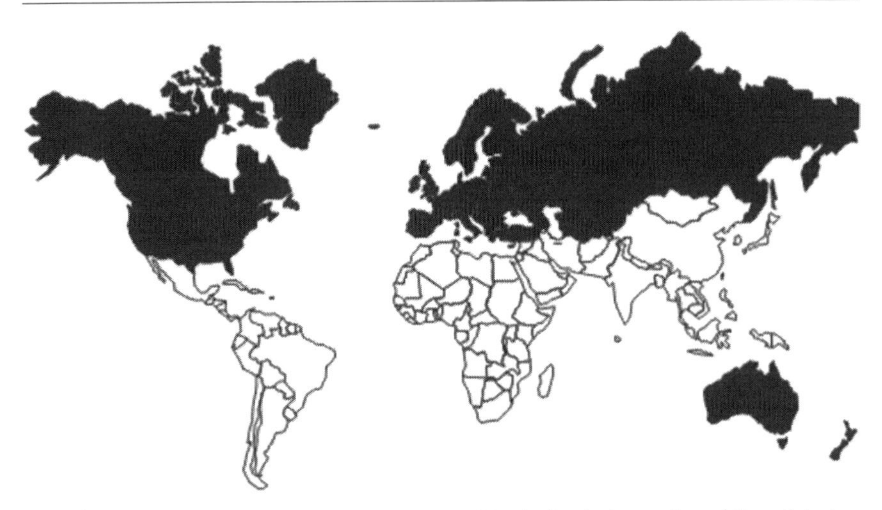

Figure 3.2 World map showing the global North (in dark grey) and the global South (in white)

Note: The allocation of countries to North or South is based on the report of the Independent Commission on International Development Issues (1980), chaired by former German chancellor Willy Brandt. The report is also known as the Brandt Report, and was instrumental in popularizing the North–South terminology in a range of disciplines.

1980s, socio-economic differences between countries have often been framed in terms of a geographical North–South divide, distinguishing the highly industrialized and affluent nations of the global North from the economically marginalized and dependent, postcolonial nations of the global South (Chant and Mcllwaine 2009; see also Figure 3.2).

However, the North–South terminology is not unproblematic. While Euro-America is the uncontested core of the North, what about the countries of the former Soviet Union? Are they North in the same way? And what about Australia and New Zealand, which are located in the southern hemisphere? They share with the broader South a sense of being on the periphery, removed from the Euro-American centers of power (Connell 2007). However, Australia and New Zealand also share many characteristics with countries in the North. They are wealthy societies with large settler populations of European origin, and daily life is very different from that in the rest of the South. Another odd one out is China, which wasn't exploited by Europeans in the same way as the prototypical South of Africa, South America, the Caribbean and large parts of Asia. The North–South nomenclature also obscures important differences between countries of the South. Thus, Singapore, India and Sierra Leone are all located in the South and share histories of (British) colonization and exploitation. However, their current socio-economic status and realities of daily life could not be more different. In 2012, the life expectancy at birth exceeded 80 years in Singapore, compared to 66 years in India and 48 years in Sierra Leone;

mean years of schooling were 10, 4.4 and 3.3 respectively; and the annual income was US$52,613, US$3,285 and US$881.[5] Jean and John Comaroff (2012) have argued that the global South is as much a geopolitical concept as it is a relational one. The South is defined not only by what it is (postcolonial and located in the southern hemisphere), but also by what it isn't, that is, powerful and affluent. In this sense we can speak about the South in the North (e.g. impoverished ghettos in the urban centers of North America) and the North in the South (e.g. affluent suburbs and private schools in India).

Hard-edged socio-economic realities are an important aspect of North–South differences, and they matter when it comes to whether people can afford to invest in digital technology. Yet it is not all about economic resources, and use of technology is also shaped by educational background, that is, general literacy as well as computer literacy skills (Van Dijk 2013). And just like access to technology, access to educational opportunities is unequally distributed across the world.

The United Nations' Human Development Index (HDI) is a useful measure in this context. The HDI includes – in addition to health (life expectancy at birth) – data on prosperity (per capita income) and education (mean years of schooling). Four types of countries are distinguished in the annual reports: those with very high, high, medium and low development. Very high development is attested for the North (Europe, parts of the former Eastern bloc, North America, Australia, New Zealand and Japan), as well as a handful of countries in the global South (Argentina, Chile, Korea, Singapore, Hong Kong, Qatar, Brunei, the United Arab Emirates and Israel). Most of the remaining countries of the former Eastern bloc, large parts of South America and most of the Arab world (including North Africa) have a high development index. Medium development is attested for some African countries (such as South Africa, Namibia, Botswana, Ghana and Equatorial Guinea) and the remaining South American countries, as well as much of Asia (including China) and the Pacific. Low development – the prototypical South – is largely an African affair, and three-quarters of countries in this category are on the African continent. Other low-development countries include Papua New Guinea, Pakistan and Bangladesh, as well as Haiti and Afghanistan (UNDP 2013). To return to the example above: Singapore is classified as a country with very high development, India as medium development and Sierra Leone as low development. On the basis of the HDI, we might want to redefine the global South broadly as referring to all those postcolonial countries located in the southern hemisphere that do not show very high development. However, this means that some postcolonial countries of the South – including Singapore, Argentinia, Chile and Barbados – fall out of the southern category, in the same way as Australia, discussed above.

The Human Development Index correlates closely with the ICT Development Index (IDI), published by the International Communications Union, a United Nations agency (ITU 2013). Again, African countries – with few exceptions – are found in the bottom quartile. However, the match is not complete and there are exceptions. For example, Kenya, a country classified as showing low development, is doing well in terms of digital connectivity. India, on the other hand, shows the

Figure 3.3 The world's least connected countries in 2012 (in dark grey; ITU 2013)

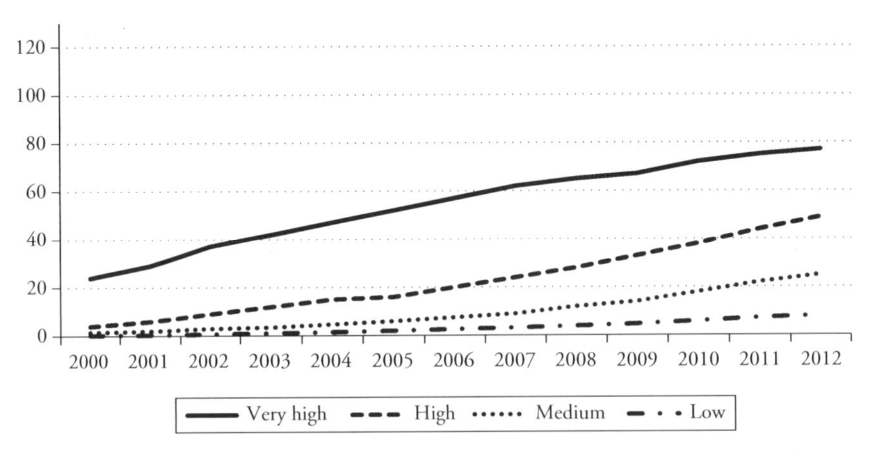

Figure 3.4 Percentage of individuals using the internet, 2000–12, grouped according to the Human Development Index (ITU)

opposite: like South Africa and China it is in the medium development category, but it remains among the group of the least connected countries (LCCs; Figure 3.3).[6]

The diffusion of computers and laptops remains low in many of the world's poorer countries, including those in the medium development category (Figure 3.4). Mobile phones, on the other hand, have shown dramatic growth since 2005. Then, there were about 2 billion mobile phone subscriptions worldwide; in 2013, less than ten years later, this had risen to close to 7 billion (for a global population of approximately 7.1 billion; Figure 3.5).

It is important not to interpret Figures 3.4 and 3.5 in ways that reproduce a

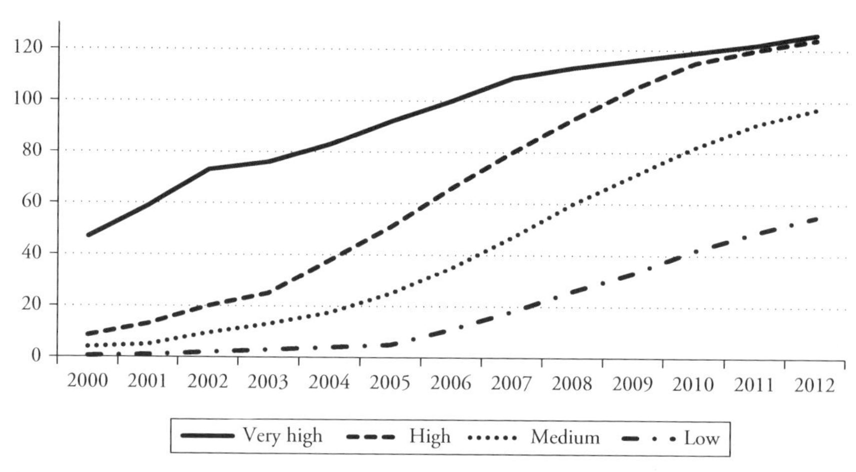

Figure 3.5 Mobile phone subscriptions per 100 inhabitants, 2000–12, grouped according to the Human Development Index (ITU)

Note: ITU statistics provide data only on mobile phone subscriptions, not on users or handsets. Thus, one user might have purchased different SIM cards for different devices (phone, laptop, tablet), or use different cards to exploit various benefits (such as free calls on the weekend from one provider, and free evening calls from another). In addition, some of the SIM cards may be inactive.

meta-narrative of modernization and development, depicting the countries of the South as playing catch-up with the North. We do not live in different temporalities with one part of the world modern and the rest of the world trying to become modern, following the same path and aspiring to the same outcome. Rather, 'we all live equally in the present' (D. Miller 2011: 16). That is, we all live in a world that holds the promise and potential of digital connectivity. Yet the way we engage with this promise differs from locale to locale. In order to understand what it means to be connected, we need to approach the South as a co-producer of what connectivity might mean and what it looks like.

Access, for example, is only part of the story for many users in the South. Slow internet connections across Africa make it cumbersome to view and upload video material – even for the local elites who live a life of plenty. In 2012, South Africa operated on 18,700 bits/s per internet user, compared to the United Kingdom's 188,875 bits/s (ITU 2013). Other countries have even slower connections, such as Mozambique with 1685 bits/s and Angola with 586 bits/s. In addition, data costs are high compared to those in the North. This has direct consequences for content production and consumption. For example, users often disable images in order to save money and time, leading to a strong preference for text-based practices.

Moreover, the experience of connectivity in the South is strongly shaped by mobile-centric access; that is, for many users in the South digital communication, from texting to email and the internet, takes place on the screens of their mobile phones. These phones are not smartphones with fairly large screens and multiple applications; rather, mobile access in the South is commonly via so-called feature phones. These mid-range phones allow for internet access, while basic mobile phones have only call and text capabilities. In South Africa, a simple WAP-enabled feature phone could be purchased for around US$50 in 2012.[7] This is not cheap, but it is becoming affordable for ever more people, and can provide internet access to those who can't even dream of buying a computer or a smartphone. Jack Linchuan Qiu (2009: 12) describes a similar situation for China. As mobile phones have become less expensive, 'hundreds and millions of have-less people have gained [digital] access through low-end ICTs'. Only recently, with the introduction of smartphones, has the North begun to catch up with the South in terms of mobile-centric internet access. Thus, although phones have been widely available in the North since the 1990s, their use was generally limited to calls and text messages, and online browsing was done on computers. Only Japan and Korea have a long tradition of mostly mobile access in a context where computers were accessible (Ishii 2004; Takahashi 2010).

In addition to inequalities between countries, there also exist access differentials within countries. National and regional averages often hide intra-national inequalities, both in the North and in the South. Socio-economic class plays a role across the world, as does the urban–rural divide. Those with less income and those residing in rural areas tend to be less connected than their middle-class, urban counterparts. This is as true for the United States as it is for Mozambique or Indonesia (Robinson and Crenshaw 2010). Other variables have changed with time in some countries, but not in others. Gender differences, for example, were observed in Europe and North America in the 1990s, but are leveling out. The situation is different elsewhere and a gender gap still exists for many African and Asian countries (Gillwald et al. 2010; Zainudeen et al. 2010). The gender gap is not simply a question of owners vs. non-owners. In rural Kenya, for example, women frequently own phones, but these tend to be of a lower quality than those of men, and women have less disposable income than men to purchase phone credit. This means that even though they have access to the device itself, they generally can *do* less with the technology (Murphey and Priebe 2011). Design is another issue that leads to inequalities despite access. Current designs tend to disadvantage older people as well as people with disabilities. The small buttons on most phones require nimble, well-coordinated fingers, and the increasingly popular touch-screen keyboards cause difficulties for people with impaired vision.

An important aspect of ICT use, which is rarely captured by available statistical data, is the importance of shared access; that is, using a phone that belongs to a friend, neighbor or family member. In Pakistan, for example, those who cannot afford a phone are not entirely cut off from telecommunications, as it is usually possible to borrow someone else's phone (Sivapragasam and Kang 2011; Ureta 2008, for Chile; Porter et al. 2012, for Ghana, Malawi and South Africa; Tacchi et al. 2012, for India). And even in cases where private phone ownership is widespread, sharing can be used to obtain access to specific features, such as music, games, a camera or the internet. Sharing arrangements are complex and locally differentiated. Sometimes sharing one's phone with others is an expected social norm (Smith 2006, for middle-class Nigeria); at other times it occurs, but is done only reluctantly and can even be refused (Sey 2011, for, especially, urban users in Ghana; Burrell 2010, for Uganda). While sharing gives access, it does not allow for anytime/anywhere communication, and thus produces an experience that is quite different from that of private ownership. Sharing is often a make-do strategy that is born out of necessity; however, it can also be used to strengthen social bonds and signal trust and intimacy. Alexandra Weilenmann and Catrine Larsson (2002) discuss examples of friendship-based sharing practices among Swedish teenagers, and Amanda Lenhart and her colleagues (2010) report that about a quarter of teenagers in the United States regularly share phones with parents or siblings. Thus, intimate sharing is not limited to the have-nots and have-less people.

In the next two sections I will look at some aspects of how people use digital technologies, first in contexts of affluence, and then in contexts of scarcity. What are the typical forms of participation in societies when resources are plentiful (the prototypical North)? And what happens when the budget is tight and one has to manage digital engagement on a shoestring (the South)?

WHEN RESOURCES ARE PLENTIFUL: GENRES OF PARTICIPATION

The United States is among the top nations when it comes to per capita internet users, and, as elsewhere in the world, the demographic group most commonly online is that of teenagers and young adults. The association between young people and technology is strong in the popular imagination, and researchers too have often focused on the digital youth, who have been referred to as – among many other terms – the net generation (Tapscott 1998) or digital natives (Prensky 2001, 2011). Irrespective of the term chosen, the basic idea is that those born after 1980 – the first generation to have grown up with the internet – have a different relationship with communication technologies than those who encountered them later in life. Some scholars even distinguish a second generation, namely those who have grown up with the participatory structures of Web 2.0 technologies, as opposed to those who grew up with the read-only internet (Helsper and Eynon 2010).

It is not unreasonable to assume that growing up with a particular technology might support specific ways of relating to it; that is, taking it for granted rather than marveling at its capabilities, or feeling intimidated by it. However, exposure to, and mastery of, technology are not simply a matter of generation. They are also about opportunities for engagement, and there are many who grew up without digital technologies but have adopted them enthusiastically. Moreover, not everyone born after 1980 has 'grown up digital'. This is particularly true for the South, but also applies to the North, where access, as discussed above, is shaped by demographic factors such as gender, class, place of residence (urban/rural) and ethnicity/race (Brown and Czerniewicz 2010; ITU 2013).

The Digital Youth Project, which was carried out by Mizuko Ito and her colleagues between 2005 and 2008, provides an ethnographic perspective on how young people in the United States have integrated new media into their daily lives and their out-of-school literacy practices. The focus of the researchers' analysis was on the social activities young people carry out using technology and the things they do when they are online. The study identifies three main types of everyday media engagement: hanging out, messing around and geeking out. Ito and her colleagues refer to them as 'genres of participation', thus emphasizing their social and conventionalized nature.

Hanging out is by far the most common type of digital engagement, and new media have become important spaces of conviviality, for keeping in touch with friends and maintaining ongoing social contact (see also Chapter 8). Teenagers in the United States, and elsewhere, spend much of their time on social network(ing) sites, and engage in instant messaging and texting.[7] Contrary to much popular opinion, such hanging-out practices are not trivial. Hanging out successfully requires considerable writing skills (further discussed in Chapters 6 and 7), and activities such as profile creation, music uploads, casual gaming or photo sharing move teenagers from digital *interaction* toward media *production* (Ito et al. 2010: 255ff.). The construction of an online profile, for example, can be a time-consuming activity. After providing basic information about sex, age and location, there is the option to select a profile and background picture, to send invitations to potential friends, to complete the 'about me' feature and to add favorite quotations, movies, books etc. Sometimes the presentation of the self is carefully orchestrated; at other times, the process is more casual, reflecting playful copy-and-paste activities. Carlos, 17 years old, explains the work involved in designing his MySpace profile page as follows:

> I just go to a certain website and if it looks like it has a lot of funny stuff I just go through that whole page and if I find something I like I just copy, paste it and put it there [on my profile]. (Ito et al. 2010: 260)

Carlos' copy-and-paste activities, his web cruising in search of interesting content, move digital engagement toward the next genre: messing around, a more intense engagement that goes beyond sociability and involves learning how to do a range of things online.

In order to understand how people engage with contemporary media, Sherry Turkle (1995) draws on Charles Lévi-Strauss' ideas of tinkering and bricolage. The two concepts describe a practical rather than abstract mode of thinking and learning, which has become important for digital engagement. In the past, it was common for people to attend computer classes in order to learn about technology. Thus, while still at school in the 1980s, I remember classes at a local computer centrum where we were taught basic programming as part of the math curriculum. I also recall attending university training in the early 1990s to learn how to use email and the internet. Such classes have all but disappeared. Today most people start by doing: they improvise and learn as they go along, and if they get stuck they ask a friend, not a teacher.

As described by Carlos, an important starting point for messing-around practices is the use of search engines, to explore the information offered, to move from link to link, to gradually refine search terms and to develop strategies for locating information online. Other examples include image or video editing, maintaining a blog, getting interested in gaming or designing a simple personal webpage. Messing around is strongly facilitated by the *type of access* a person has. It is easiest 'when kids have *consistent, high-speed Internet access*, when they *own gadgets* such as MP3 players and DVD burners, and when they have a great deal of *free time, private space, and autonomy*' (Ito et al. 2010: 61, my emphases). Messing-around practices are fairly common in the North, but rarer in the South, where – as discussed above – the necessary high-end technology is rarely available, data speed remains low, data costs are high, and the daily life of many does not include private space and ample free time.

Finally, geeking out refers to forms of intense and focused media engagement, typically reflecting highly specialized knowledges that have been accumulated over time. Although largely driven by one's personal passion, geeking-out practices are not solitary, but embedded in networks of sharing and collaboration. Moreover, unlike the ubiquitous hanging-out practices, geeking-out networks tend to move beyond the local circle of friends, and bring one into contact with people from a wide range of backgrounds and countries. Examples include intensive gaming, especially massively multi-player online games, the production and sharing of content linked to various forms of media fandom, Wikipedia editorship and regular text or video blogging. As in the case of messing around, consistent and high-quality technology access is essential. Equally important, however, is the existence of peer-learning networks and sufficient time to improve one's skills. In the literature such learning networks are referred to as 'communities of practice' (following Lave and Wenger 1991) or as 'affinity spaces' (following Gee 2005): people come together around common practices, interests and activities, and develop a sense of community based on like-mindedness.

The three genres of participation identified by Ito and her colleagues allow us to capture digital engagement in broad brushstrokes, and provide a vocabulary for describing the different ways in which teenagers, and adults, engage with new media. Importantly these genres draw attention to the significance of easy access to high-end technology, ideally at home. Public access, on the other hand – that is, through schools, libraries or internet cafés – is far from optimal for digital engagement. The time spent on the computer is usually limited in order to ensure that everyone gets a chance. Moreover, costs can be an issue; there are strict closing times; users lack control over network settings, and certain websites might be blocked; hardware is not always well maintained; software might be outdated; and finally, there are concerns about privacy and data storage. Plentiful around-the-clock and high-speed home access is generally seen as the 'gold standard' for intense digital engagement (Warschauer and Mtuchniak 2010; d'Haenens and Ogan 2013). Easy access, however, is something that is notoriously absent for those living in socio-economically precarious circumstances – whether within affluent societies or, and even more so, in countries of the global South. The next section explores the types of digital practices we see when resources are scarce and personalized online access is limited to fairly basic feature phones.

WHEN RESOURCES ARE SCARCE: COMMUNICATION ON A SHOESTRING

Mobile phones have become everyday technological objects from Accra to Mumbai, from Santiago to Kingston, from Kuala Lumpur to Jakarta. The broad diffusion of mobile phones across socio-economic boundaries is reflected aptly in the title of a poem by the Zimbabwean author Mihla Sitsha, *Makhalemkhukhwini*, 'that which even rings in a shack' (Nkomo and Khumalo 2012). Cara Wallis (2011) gives the example of Wu Daiyu, a rural migrant in Beijing, who spent almost her entire monthly salary on a simple clamshell phone. When it was stolen, she bought a new phone almost immediately. Wu Daiyu relies on her mobile phone to contact her family in the village, and it allows her to maintain important social ties. For Wu Daiyu and many others access to a mobile phone, costs notwithstanding, is not a luxury, but a necessity (Wallis 2011).

Staying in touch is important, emotionally and socially. In South Africa, in the past, rural–urban migrants who did not stay in touch with those back home were called *itshipa* in isiXhosa. *Itshipa* means 'absconder'. This is a borrowing from English 'cheap' and refers to someone who has broken his or her social ties. However, with low salaries, large distances between city and village and lack of access to telecommunications, 'to abscond' was rarely a free choice. Today it is easier to stay in touch, and in the evenings, on the streets of Cape Town, one overhears workers who, after a long day, check with people back home in the village using their mobile phones. Many are mothers, who inquire after, and speak to, the children they have left behind with grandmothers or aunts. In her work on transnational Filipino families, Cecilia Uy-Tioco (2007) describes such mediated interactions between mothers and their children as 'remote mothering'. And fathers worry too. Zodwa, a young woman who came to Johannesburg in search of work, told me how she receives regular calls from her father: 'My father calls me four or five times a week, he likes to check on me, he also wants to find out whether I have eaten.'

However, obtaining a phone is only the first step; one also needs phone *credit* (also referred to as *airtime, units, minutes or top-up*). And this can put a strain on resources. South African shack dwellers, for example, were found to use around a third of their meager income to finance their communicative needs, sometimes buying airtime rather than other essential items (Wasserman 2011). While text messages provide a fairly cost-effective way of staying in touch, many, especially older and less educated, people prefer voice calls, which allow them to get an answer quickly and do not require much in terms of literacy skills. However, because of the high costs for calls it is essential to keep them short. This economic imperative can lead to changes in conversational norms when compared to body-to-body interactions. In isiXhosa, for example, spoken greetings typically flout the Gricean maxims of quantity and manner. In body-to-body interactions, speakers do not merely greet the person they meet, but engage in a prolonged interaction that is deliberately expansive. In mobile phone conversations this has changed: phone interactions tend to be short, salutations are exchanged quickly, questions about well-being can remain unanswered and participants move quickly to the 'business of talk' (Kaschula and Mostert 2009).

The cheapest way to communicate is to exploit the zero-cost affordances offered by mobile phones. In *The Rules of Beeping*, Jonathan Donner (2007) uses data from Rwanda to unpack an innovative, zero-cost communication practice that can be used when credit is low, but one wishes to stay in touch with friends or family. Beeping (also called *flashing, buzzing, missed calling, or pinchar* 'to prick' in South America) is simple: you call a mobile phone and intentionally hang up before the other person can pick up the call. Your number or name will be visible on the recipient's phone. The cost is zero to both parties but some form of contact has been established. What does it mean to receive a missed call? Beeping can mean different things. It can be a request for a call-back; it can also be a signal that someone is thinking of you; or it can express a pre-negotiated message ('I am outside!'). Sometimes people use repetition to distinguish different meanings in ways familiar from Morse code. One missed call might simply mean 'I am thinking of you', two (or more) might mean 'something happened, I don't have airtime, call me'. According to Donner, there are four basic maxims that guide beeping practices.[8]

> *Maxim 1:* You can beep people with more money (the principle of 'the richer guy pays').
> *Maxim 2:* You can beep most family and close friends if you have run out of airtime and would like to speak to them. They will usually call you back if they have credit.
> *Maxim 3:* If you want to make a good impression on someone, don't beep them.
> *Maxim 4:* And always remember, do not beep too much.

Maxims 3 and 4 work together: too much beeping can be a problem, especially in business and courtship contexts when impressions need to be carefully managed. Heather Horst and Daniel Miller (2006: 120) quote a young Jamaican woman who had little time for her beeping suitor:

> I think it was really cheap of him, if you want to talk to me, buy some credit simple as that. Basically, I just don't like the idea, and sometimes it's not even important when you do call them back, 'me just call fi see if you alright'. Cho!

The very fact that mediated talk is not free allows us to gauge the other person's commitment. If someone is really interested in you, they will invest not only time and effort, but also money, in talking to you.

Realizing the importance of zero-cost practices, some operators have introduced please-call-me (PCM) messages. Unlike missed calls, these PCM messages generate advertising income for the network. Instead of beeping someone, PCMs allow one to send a free text message. The message says 'please call me' and the user can insert a small number of free characters. The message is followed by one's phone number and a short advertisement. The intention is that users should insert their names after 'please call me', thus creating standardized minimal texts such as: 'Please call me ana [my mobile phone number] [advertisement]'. However, it did not take long for users to appropriate this new option in ways that were not intended by those who invented the technology. In their work on Jamaica, Horst and Miller (2006) found that sending call-back messages was a daily digital practice, and that for those living in low-income communities PCMs had become 'almost synonymous with texting' (p. 70). The free characters are used not to insert one's name (as intended), but to include an additional message (as the number of the sender is visible anyway). In example (3), Mame, who lives in Quthing, Lesotho, informs her family that she has arrived home safely using a free PCM. Instead of using the free characters to write her name, she inserted *ke fihlile* (Sesotho, 'I have arrived').

(3) Please Call ke fihlile [phone number] Vodacom customers can now send 10 FREE Please Call Me every day & you can personalise it once daily. (Lekhanya 2013)

Mobile phones certainly provide affordable person-to-person connectivity, and calls and text messages are a ubiquitous practice across the world. Things, however, are different when it comes to internet access. Not everyone who owns a mobile phone has access to the internet. Polo Lemphane and Mastin Prinsloo (2013) discuss the Mahlale family in Cape Town. The two parents, both unemployed and dependent on government grants, and their five children, aged five to fourteen, live in a makeshift shack. Like most South Africans, the parents own phones, yet these are of the most basic type and don't allow internet access. The Mahlale family belongs to the roughly two-thirds of South Africans who are not connected to the internet. The same is true for Mame in example (3). She texts and calls regularly, but her phone does not have internet access. In South Africa, those not connected to the internet share certain characteristics: they tend to be older, be unemployed and struggle to read and write, and many live in the rural areas of the country.

South Africa, like many other economies of the South, is characterized by high income inequality and unequal life chances for those growing up; it is a land of plenty as well as a land of scarcity. In our research, we collected comparative data from students attending two schools in Cape Town: a government school located in a low-income working-class neighborhood, and a private school in a middle-class neighborhood. Annual fees at the private school are more than 120 times higher than those at the government school (US$5,700 and US$45, respectively). The private school boasts well-kept sports fields and science facilities, while the government school is run down and marred by gang violence, which is rampant in the area.

Almost all students at the two schools own phones, and teenagers at the government school actually begin using phones slightly earlier (Table 3.1). This is the familiar story of near-universal mobile phone access, crossing the lines of class and race. Things look different once personal computers and high-end smartphones are considered as part of the mix. Whereas over 90 percent of students at the private school own computers (desktops and laptops combined), the figure is only around

Table 3.1 ICT access and use at two South African schools (2011; N = 240; 20% sample of full population for both schools; age range 13–17).

		Government school (in percentages)	Private school (in percentages)
Access	Owns desktop	22.8	49.5
	Owns laptop	13.8	47.4
	Owns mobile phone	84.8	97.9
	Owns smartphone	8.4	51.6
	On contract	7.6	49.5
	Uses less than $4 airtime per week	58.0	31.6
	Has used a phone by age 7	42.4	34.9
	Has used a computer by age 7	33.2	70.5
Use	Never searches the internet	9.3	2.1
	Searches the internet mainly via phone	43.1	11.6
	Does not send emails	60.7	21.3
	Sends emails mainly via phone	22.1	10.6
	Uses Mxit	91.0	76.7
	Uses Facebook	42.1	86.3
	Uses Twitter	11.7	34.7
	Uses other social network	11.7	26.3

one-third for the government school. Moreover, middle-class students attending the private school were exposed to computers – which afford users a greater variety of digital practices than phones – at a younger age than those attending the government school. With regard to costs, more students at the private school are 'on contract' (as opposed to pre-paid pay-as-you-go access). This means that they benefit from lower call and data rates, and usually have parents settle their bills (as contracts can only be taken out by credit-worthy adults). Following Indra De Lanerolle's (2012) typology of South African media users, we can describe the students at the private school as *superconnected*: many of them have access anywhere and anytime, just like their counterparts in the North. Students at the government school, on the other hand, rely on their phones and public facilities (libraries, internet cafés) for online access. They are *connected*, but not superconnected.

Differences in material access support differences in use. Messing-around practices (searching the internet, trying out new sites, sending emails) are, on the whole, more common for students at the private school. However, when it comes to engaging in such practices from a phone, students at the government school are trendsetters – out of necessity rather than choice. Hanging-out practices, on the other hand, are common in both groups. However, the teenagers hang out in different online spaces and their choice of platform reflects, yet again, their access to technology. Students at the government school make ample use of Mxit. This is a South African mobile instant messaging service that was designed especially for feature phones. It allows users to maintain a simple profile and to engage in one-to-one chats as well as group chats. Mxit allows convenient access of sort, but has none of the 'bells and whistles' of internet-based applications, such as Facebook, which is popular at the private school. Although Facebook can be accessed via mobile phones, profile set-up, including uploading photos and other material, is best done via desktops. Moreover, data costs can be prohibitive for low-income users as Facebook images, especially, are high-resolution and thus use up airtime or credit. In the South African context (and at the time when the data was collected), using Facebook, rather than Mxit, was a form of conspicuous consumption: it indicated one's position in the world of digital interactions, one's access and affluence.

CONCLUSION: THE TIMES ARE A-CHANGIN'

Like all communicative resources, digital technologies are'placed resources': their use is shaped by local contexts, needs, practices, the material conditions of the everyday and the larger political economy (Prinsloo 2005). The overall situation is shaped by persistent inequalities, but is also dynamic as globally more and more people gain access to mobile communication technologies and the internet.

Although current Web 2.0 environments encourage content creation, different forms of access shape user engagement. Simple feature phones are not optimal for digital creation because screens are small, applications for editing are limited, connection speed is slow and data costs are high. As noted by Katie Brown and her colleagues (2011: 145): 'Mobile internet is better than no internet but cannot match the interface of traditional computers.' Smartphones might be a game-changer. Their capabilities are closer to those of a computer: they allow for easy multimedia consumption and production, fast internet access and feature large screens. However, although smartphone sales are growing worldwide, this is unlikely to change the global face of mobile access anytime soon. In the global South, old, basic phones are not discarded but passed on as gifts, and there is an entire industry that specializes in recycling and reselling old handsets. It is more than likely that low-end call/text-only phones and feature phones will be around for a while.

Yet even on shoestring budgets and with less than optimal technology, people make the most of what is available to them, and the diversity of digital engagement in the South should not be underestimated. Users in the South do not only hang out, they also mess around with the technology they have. For example, they might edit photos on their phones, Bluetooth the latest music to one another, and use Google and Wikipedia to access information. If they need access to specific features – such as uploading images or videos, reading longer texts or doing more complex layouts – they might look for a public computer, in an internet café or a library. Users also produce copious amounts of writing on mobile phones. Zodwa, an 18-year-old South African student, posted 228 status updates (in English and isiXhosa) during her last year of school (2010), amounting to over 4,000 words, equivalent to a medium-length university essay. With a few exceptions all posts were sent from her feature phone (while I wrote only about 1,000 words in the same year from a computer). And plentiful access notwithstanding, not everyone in the North is geeking *out*, or even *messing around*. For many, digital engagement remains limited in scope to some hanging out on social networking sites and lots of read-web; that is, the retrieval and consumption – but not creation – of content (ITU 2013).

NOTES

1. Spoken language too is affected by the political economy, albeit in less direct and tangible ways. Thus, certain forms – typically those associated with upper- or middle-class speakers – are valued more highly than others, creating persistent structures of linguistic inequality and disadvantage. Similarly, some languages are valued more highly than others (Gal 1989; Bourdieu 1991; see also Chapter 4).
2. TMB ('text message break-up') is a topic well represented on YouTube, where video blogs and comedy skits portray the social unacceptability of breaking up by text.
3. *Gyandoot* is Hindi and means 'purveyor of knowledge'.
4. danah boyd and Nicole Ellison (2007) have argued that although Facebook and other applications allow for 'network*ing*', they are primarily used by people to 'articulate and make visible their [existing] social networks'. These authors consequently favor the term 'social network site' over 'social networking site'. I will use the combined term 'social network(ing) site' to indicate that participants use these applications for both network maintenance and network formation.
5. The global and regional statistics come either from the United Nations Development Program (www.undp.org) or from the International Telecommunications Union (www.itu.int). They are, unless otherwise indicated, for 2013.
6. The IDI combines several indicators of *access* (such as mobile phone subscriptions, bandwidth, percentage of households with computers), *usage* (percentage of individuals using the internet, through either mobile or fixed line subscriptions) and *skills* (literacy rate, enrolment in secondary and tertiary education). Unlike the HDI, which has been published annually since 1990, the IDI was first published in 2011.
7. Popular feature phones are the Nokia 5130 XpressMusic, Nokia 2700c and Samsung C5212. A prototypical example of a basic phone is the Nokia 100.
8. Donner calls them 'rules', which implies a greater degree of normativity than seems warranted.

Chapter 4

Virtual landscapes: practices and ideologies

Created by User:Tatiraju.rishabh/userboxes/multilingualism

INTRODUCTION: MULTILINGUALISM ONLINE

Historically, the internet was an English-dominant space. Not only did most of the technological development take place in the United States, but this was also the country where its early users were found. However, steady growth in global access, as discussed in Chapter 3, has meant that ever more people – from different linguistic, social and geographic backgrounds – have gained access to communication technologies, and as a result the internet has moved beyond 'the English milieu of its birth' (Prado 2012: 47). The internet's growing linguistic diversity was also supported by technological developments. Especially important was the move from ASCII – which included only Latin letters – to Unicode, which allows access to a wide range of scripts. And with the rise of multimodality, sign languages and languages without a written tradition could be represented too.

The American linguist David Harrison[1] has called this the 'flipside of globalization': rather than inevitably assimilating to a dominant – typically English-speaking – global digital culture, people use digital media to express their voices and linguistic-cultural practices (Danet and Herring 2007; Ginsburg 2008). At the same time, old hegemonies have not disappeared. Although webpages are available in different languages and social media platforms have created new opportunities for multilingual interaction, certain languages continue to dominate.

Languages online are represented on screens; they become visible and are on display. The area of sociolinguistics that is most overtly concerned with the representation of language and other semiotic resources is linguistic landscaping. Linguistic landscaping looks at the texts – signs, posters, websites – that institutions and people produce; at the ways in which these texts are positioned within particular physical environments, as well as fields of inequality and power; and at how the texts shape the world we live in, are seen and attended to, or overlooked and ignored by audiences (Shohamy and Gorter 2009; see also Scollon and Scollon 2003; Blommaert 2013). The concept of representation is important here. As Stuart Hall (1997) reminds us, representation is not only about reflecting – more or less accurately – an existing state of affairs in a different medium, but is also constitutive of reality. The way we experience language in the public sphere shapes not only our understandings of what is normal or unusual in such spaces, but also our ideas of what languages are and what they 'look like'.

This chapter approaches online multilingualism from the perspective of linguistic landscaping. The first section discusses representations of online multilingualism, both at a global level (the internet in its entirety) and at a local level (specific webpages). In exploring the internet as a linguistic landscape, the discussion then turns to a specific digital environment: Wikipedia, the popular online encyclopedia, which is not only free and collaborative, but also multilingual.[2] The very idea of Wikipedia – creating a compendium of knowledge in many languages for many people – means that questions of representation are paramount. The initial focus of the discussion is on the history of Wikipedia's multilingual logo as an example of ideological practice. The logo is a visible sign on every page and forms part of Wikipedia's stable, and indeed branded, semiotic landscape. The second part of the chapter looks at multilingual editorial practices with particular attention to the isiXhosa Wikipedia, one of the many small editions on the site. IsiXhosa, one of South Africa's official languages with about eight million speakers, is present online, yet its digital use remains restricted.

VIRTUAL LINGUISTIC LANDSCAPES: GLOBAL SPACES AND LOCAL PLACES

Dejan Ivkovic and Heather Lotherington (2009) have described the internet as a virtual linguistic landscape, thus broadening the linguistic landscaping paradigm, which is grounded in physical geography, to include the representation of language(s) in publicly accessible digital spaces. These authors argue that the use of languages in digital contexts is an essential part of the 'global linguistic ecology', and that 'fragile balances in individual and social repertoires' (p. 32) are affected not only by linguistic choices made offline, but increasingly also by choices made online. This means that an interest in studying virtual linguistic landscapes intersects with sociolinguistic work on language choice, shift/loss and maintenance. Writing about the possibility of 'reversing language shift', Joshua Fishman (2001) placed considerable emphasis on literacy as a stabilizing factor. Although he focused on traditional print literacies, his argument can be extended to digital contexts. That is, non-existent or limited online presence might negatively affect language maintenance and accelerate processes of language shift. The importance of new media is made more explicit in the UNESCO document *Language Vitality and Endangerment* (2003), which emphasizes that the way in which speakers respond to new media is important for the continued use of a language in the twenty-first century.

What kind of linguistic landscape is the internet? What languages are visible and vibrant online, and what languages are invisible? Or – to rephrase Jan Blommaert and his colleagues (2005: 198) – how does the internet organize 'particular regime[s] of language', which 'incapacitate' those who are not literate in the languages on display? The initial focus of the discussion is quantitative, taking the internet as a corpus and each website as a sign, and aims to estimate the relative distribution of languages online. Given the size of the internet – close to two billion indexed webpages (March 2014; worldwidewebsize.com) – the seemingly straightforward process of counting the languages used on different websites is methodologically challenging, and researchers need to develop ways of dealing with massive data sets (what is called 'big data'; boyd and Crawford 2012).

Daniel Pimienta and his colleagues (2009) have spearheaded a simple, yet useful, method for determining the web-presence of specific languages. First, a set of keywords is identified and translated into the languages under investigation.[3] Each keyword is then entered into a search engine, which provides frequency counts for the word's occurrence. Finally, an average score is calculated for each language. This is a neat methodology. However, it suffers from problems with search engine behaviour and the results produce trends rather than reliable absolute numbers.

The approach can be illustrated with a simple Google and Yahoo! search for one item from Piementa's list of keywords: the adverb *today* in four languages of my own repertoire, namely English, German (*heute*), Afrikaans (*vandag*) and isiXhosa (*namhlanje*). On Google, *today* shows up on thousands of millions of webpages and *heute* scores more than one hundred million; even *vandag* reaches five million, and only isiXhosa trails behind with forty-three thousand (Table 4.1). The results indicate a clear hierarchy: English is far ahead, followed by German, Afrikaans and finally isiXhosa. This hierarchy is the same on Yahoo!, but actual numbers are considerably lower; isiXhosa, for example, drops to a mere eight thousand hits.

Sometimes the item occurs, but the page is not actually in the language to which the word belongs. Thus, one of the first Google entries for isiXhosa *namhlanje*

Table 4.1 English, German, Afrikaans, and isiXhosa: web-presence estimation using the keywords *today, heute, vandag and namhlanje* (date of search: May 15, 2012; results are rounded)

Search engine	English	German	Afrikaans	isiXhosa
Google	6,800,000,000	110,000,000	5,000,000	43,000
Yahoo!	2,600,000,000	100,000,000	228,000	8,000

comes from the English-speaking site WikiAnswers: *What does 'namhlanje' mean in Xhosa?* Another entry directs me to *Namhlanje Travel Agency*, followed by a video for Abdullah Ibrahim's song *Siyahamba namhlanje* ('we are leaving today'). Moreover, *namhlanje* is also used in isiZulu, a closely related language, and the count thus includes pages in both languages. Similarly, for *vandag*, Yahoo! includes the Dutch spelling *vandaag* in the search, and lists sites with both spelling variants (even when preferences are set to 'search exact phrase'). Thus, while search engines are freely accessible to researchers and offer a convenient tool for the broad estimation of web-presence, the results need to be interpreted carefully. They indicate trends and patterns, rather than reliable actual numbers.

The broad, global pattern of online multilingualism is fairly stable. English remains the strongest language online, followed by a group of languages located in the global North: German, Russian, Spanish, French, Portuguese, Italian and Polish, as well as Japanese (Figure 4.1). These are fairly powerful languages, with histories of state support and spoken in regions of the world where internet access is, on the whole, easy and affordable. The only language from the global South that has a noticeable online presence is Chinese. The vast majority of the world's six thousand to seven thousand languages remain invisible or minimally represented.

Geographical areas or countries might, however, show 'regimes of language' that differ from this global pattern. In a study carried out in 2010, researchers at the Language Observatory in Tokyo collected a large sample of webpages from particular geographical domains; for example, webpages whose country identifiers – the

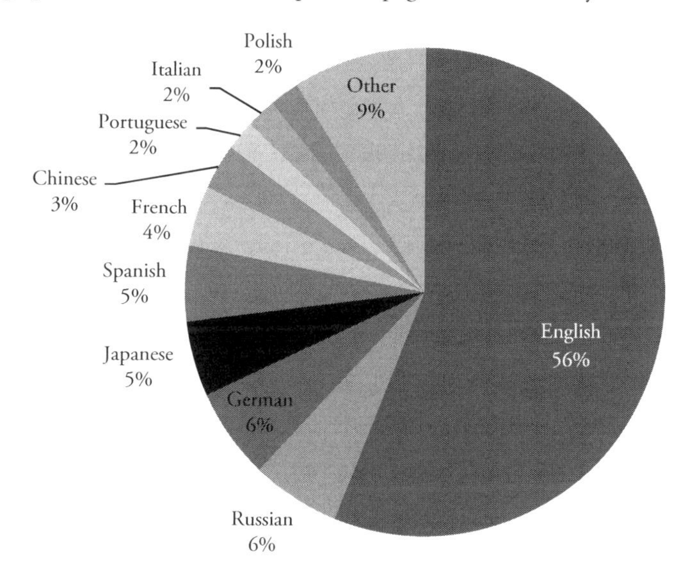

Figure 4.1 Languages used on the top ten million webpages: percentages (http://w3techs.com/technologies/overview/content_language/all, May 14, 2014)

last two letters of the URL – locate them 'in Africa'. The researchers found that African web-content was overwhelmingly in English (over 80 percent), with French a distant second (6 percent) and African languages invisible. English was also the most commonly used language on Asian webpages; however, unlike in Africa, English did not dominate and was below 50 percent. A rather different pattern was found in the Caribbean, where Spanish was the dominant language (Yoshiki and Kodama 2012).

The fact that most of the world's six thousand to seven thousand languages are absent from, or minimally represented in, virtual spaces suggests that the digital constitutes a linguistic ecology quite unlike that of spoken languages. It is more similar to what has been called the ecology of literacy, reflecting the way in which literacy practices are shaped by social and political environments (Barton 2007). For example, in many postcolonial countries, English, French or Portuguese remain the languages of literacy, education and public signage, while local languages are used mainly in spoken interactions. This offline pattern, reflecting a communicative separation between literacy and oral practices, is then reproduced and remediated in digital spaces, and languages that do not show sustained literacy use 'on paper' are less likely to achieve this 'on screen'.

Yet the link between offline and online practices is fragile, and virtual linguistic landscapes are in many ways quite unlike physical landscapes. Perhaps the most important difference is the fundamentally open, unbound and delocalized character of the internet, where anyone can enter, read, contribute and participate. And, unlike signage, which is always embedded in particular physical environments, digital content is mobile and visible on diverse screens (phones of various types, iPads and computers). These, in turn, are located in various physical contexts (living rooms, classrooms, bus stops, airports and so forth; see Chapter 2). It is also a space where the status quo of the offline literacy ecology – formal and typically linked to educational institutions and discourses – can be challenged, and the digital has become a space where existing literacy regimes are contested and subverted. Consider again isiXhosa, a language whose overall web-presence – as shown above – is very limited. Most speakers of isiXhosa are also bilingual in English, and overall English tends to dominate their online engagement. However, on Facebook groups where the name itself signals a strong cultural identity – such as *Indlu yakwaXhosa* ('the house of isiXhosa') – isiXhosa is valued and expected (Deumert and Lexander 2013; Deumert forthcoming a; see also Chapter 6).

However – and this is a methodological issue for the quantitative studies of web-presence discussed above – interactive spaces such as Facebook are usually not captured by existing search engines. It is therefore important to complement a bird's-eye view of web-presence with the careful analysis of particular spaces. A focus on specific contexts is also common in linguistic landscape research, most of which looks at particular *places*, such as cities or individual neighborhoods; that is, *spaces* to which people feel attached, where they show investment and actively shape the environment in distinctive ways. Thus, while we click ourselves through the internet quite individually, creating numerous, mostly transient, pathways, and experience different landscapes, some pages might become fixtures in our lives and we, and others like us or with similar interests, will visit them regularly. They become something akin to 'virtual neighbourhoods' (Appadurai 1996: 195), online *niches* of interaction that might display unique linguistic and semiotic landscapes (see also Chapter 8).

Examples of this are the diasporic websites studied by Jannis Androutsopoulos (2006). One of the sites was the theinder.net. Aimed at the Indian diaspora in Germany, theinder.net was launched in 2000. The domain name reflects the community's hybrid identity: *Inder*, German for 'Indian', is combined with the English definite article *the*. The name is also a pun on 'internet'/'indernet' (a wordplay that goes back to a popular German comic strip from the 1990s). The site's bilingual orientation and accessibility are visible in the site slogan (which is given first in German, then in English), and informs site architecture: the start-page is German, but provides a link to the English version (Figure 4.2). When Androutsopoulos collected his data in 2005, the site also included a rudimentary Hindi interface. Hindi symbolizes 'Indianness' more strongly than does English, which – although one of India's official languages – is also a global lingua franca. The Hindi version, however, did not survive and had disappeared by 2014; possibly because diasporic readers were rarely fluent in the language. In the 2014 design, the representation of 'Indianness' is visual rather than linguistic. An image of the Taj Mahal and of a tiger's eye are prominently displayed as part of the page header, separated from the surrounding images by being reproduced in color (multimodality will be discussed in more detail in Chapter 5).

Although the site design of theinder.net suggests the presence of two parallel – German and English – editions, the separation of languages is incomplete and the actual browsing experience is characterized by hybridity and multilingualism. For example, adverts on the German site are posted in both English ('Plan your trip to Kerala') and German (*Fit Reisen, Gesundheit und Wellness weltweit*, 'Fit travel, health and wellness worldwide'; March 2014). The English site shows even more hybridity. Not only are adverts visible in both languages, but the English page also displays German articles (on the left), and the English text is embedded in the German interface. Moreover, not all articles from the German version are translated. Thus, although English is present on this site, it remains secondary. The main language is German, that is, the language that dominates in the offline world of the Indian-German diaspora.

In 2005, the site also had a forum, which, according to the analysis by Androutsopoulos, was predominantly German. In 2014, the forum was no longer part of the site and its function was taken over by a Facebook page. While the old forum was clearly German-dominant, English is common in the Facebook group, in postings as well as interactive user comments. Very occasionally Hindi is used; for example, when writers discuss the lyrics from Bollywood movies. Such minimal and largely symbolic use of languages has also been described for other interactive digital environments. Wolfgang Sperlich (2005), for example, discusses the use of Niue,

Figure 4.2 A diasporic, bilingual website: theinder.net
(a) German version; (b) English version (theindernet.blogspot.de, May 14, 2014)

an Austronesian language, on the message board OKA-KOA. Although English dominated strongly, Niue was present and part of the site's linguistic landscape. And Vassili Rivon (2012) looks at the use of Eton, a Cameroonian language, on Facebook. The pattern is similar: French, the former colonial language and language of education and literacy, dominates, but Eton occurs in the form of emblematic code-switching.

A different situation is found on another diasporic site, germany.ru, which is aimed at Russians living in Germany. The site carries the programmatic slogan Германия по-русски, 'Germany in Russian'. Although it is possible to access a German version, the editors caution readers: *Wir bitten um Verständnis, daß der Inhalt der deutschen Version mit der Russischen manchmal nicht übereinstimmen kann. Bitte verwenden Sie falls möglich die russische Version* ('The content of the German version does not always entirely match the Russian version. Please consult the Russian version if possible. We thank you for your understanding'). Russian is prioritized not only on the site, but also on the discussion boards, which were predominantly Russian when Androutsopoulos studied the site in 2005, and remained predominantly Russian in 2014. Thus, unlike on theinder.net, usage on germany.ru does not prioritize the language of the host society, and provides a space for the sustained – rather than emblematic – use of Russian.

The remainder of this chapter will focus in more detail on a specific online place, Wikipedia, that attracts not only readers from across the world, but also a self-declared 'community' of dedicated contributors who call themselves Wikipedians (for a discussion of the English Wikipedia, see Myers 2010). The linguistic landscape generated by Wikipedia includes both top-down and bottom-up representations of language and other semiotic resources; that is, representations that are published by institutions (site architecture and overall design interface; top-down), and those generated by individual social actors (user-generated content; bottom-up; see the papers in Shohamy and Gorter 2009 for a critical discussions of these concepts). Wikipedia is important for media sociolinguistics for various reasons. The encyclopedia is a top site for global internet traffic, and has been celebrated in the literature as one of 'online multilingualism's greatest successes' (Bortzmeyer 2012: 381). Moreover, Wikipedia has taken the challenge of mobile access to heart. Since 2013 it has been possible to edit Wikipedia on mobile phones, and the Wikipedia Zero project (launched in 2012) provides free access to Wikipedia's mobile edition in several low-income countries. The site's broad global reach, combined with explicit policies of supporting multilingual content, makes Wikipedia an important space for understanding the potential and challenges for digital multilingualism.

PRACTICES AND IDEOLOGIES I: SYMBOLIZING MULTILINGUALISM

Wikipedia, established in 2001 by Jimmy Wales, supports a commons-based approach to knowledge creation and uses a wiki tool to allow for the collaborative editing of documents. The resulting texts are not only multi-authored but also transitional. That is, they can, in principle, be altered or amended by anyone at any time, and every user is a potential author and editor. Wikipedia is owned and managed by the Wikimedia Foundation (WMF), a non-profit organization. Its board of trustees also oversees other collaborative – and multilingual – projects, such as Wiktionary and Wikibooks.

Wikipedia is a highly reflexive space. Discussions about policies and editorial projects are publicly accessible and archived on Meta-Wiki, the Village pump and lists such as Foundation-l or Wikipedia-l. In addition, comprehensive and detailed statistics are generated on a regular basis.[4] Each change to an article is recorded and can be recovered by clicking the 'view history' tab. Articles also have 'talk pages' where users discuss the content of the page, negotiate differences of opinion and motivate their changes. An integral part of the linguistic landscape of Wikipedia is the logo, which is prominently displayed on the start-page (Wikipedia.org), and visible in the top left corner on every article page. The discussion in this section expands on the critical reading of Wikipedia's logo by Astrid Ensslin (2011) as an example of *ideological practice.*

When he founded Wikipedia, Wales used – in a rather jingoist gesture – the image of an American flag as a logo. Toward the end of 2001, this was replaced, in quick succession, by two ball-shaped logos, both of which displayed text in English and foregrounded the Anglo-European tradition. The first logo used an extract from a mathematical treatise by C.L. Dodgson (better known by his pen name, Lewis Carroll); the second logo an extract from Hobbes' *Leviathan* (1651).[5] In October 2002, User:Ch, the founder of the Esperanto edition, suggested thinking about a universal logo that would be 'neutral to language'.[6] A logo contest was announced and in September 2003 a new logo was selected. This logo reinterpreted the ball-shape of the earlier designs as an unfinished jigsaw puzzle. The visual metaphor emphasized the open-ended nature of knowledge and playfully invited participation (Figure 4.3). The text is no longer in one language only and the impression is one of great diversity. However, the print is small and it is impossible for the reader to identify, easily and confidently, different languages, scripts or words. It is the visual equivalent of a cacophony of incomprehensible sounds – as opposed to the polyphony (or heteroglossia) of the meaningful, that is, comprehensible presence of multiple languages (for further discussion of these concepts see Chapter 6).

In follow-up discussions of the logo, Wikipedians highlighted the visual and aesthetic aspects of the represented text. One of the recommendations – made by User:No – was to include more '*cool-looking* Indian writing systems, Mongolian, Tibetan, Georgian, Coptic, Hebrew, etc.' (my emphasis). User:No took his own recommendation to heart and created a new logo, which maintained the puzzle-ball design but represented linguistic diversity solely through different scripts (Figure 4.4). While most letters were selected to represent [v], the labiodental onset of the word *Wikipedia*, others – such as the Greek omega (Ω) – were 'chosen at random', simply because they looked 'interesting'.

Figure 4.3 Wikipedia logo: representing multilingualism I (September 2003)

User:No's design became the official logo in October 2003. However, it soon attracted criticism because some scripts were inaccurately represented. Consider the example of Hebrew, located to the bottom-right of the logo. User:No included the Hebrew letter ו (pronounced *vav*). The letter represents the sound [v], the first letter of *Wikipedia* in Hebrew script (ויקיפדיה). However, the letter on the logo is distorted (as a result of the 3-D projection) and looks much more like ר (pronounced *resh*), representing not [v], but an r-sound. This is comparable to someone mixing up *b* and *p* or *r* and *n* in the Latin alphabet. In the case of Devanagari, an Indian script, vowels are attached to consonants in various positions. Some vowels follow the consonant, others appear as diacritics below or above, and the short /i/ diacritic precedes the consonant. Thus, when transliterating /vi/, the correct symbols were used, व /v/

Figure 4.4 The puzzle-ball adaptation: representing multilingualism II (October 2003)

and ◌ि /i/ (short vowel, dependent form, i.e. the dotted circle indicates the position where the vowel is inserted). However, they are in the wrong order (the Devanagari characters are in the middle at the left side of the logo). That is, /vi/ was transcribed, following the Roman letter sequence, as /v/ plus /i/; yet the sequence in Devanagari should have been /i/ plus /v/. The correct symbol would be वि. Errors were also noted in the representation of the Japanese and Chinese scripts.[7]

Jimmy Wales responded in the *New York Times* that mistakes are part of the larger wiki project and will be corrected by contributors in time (Cohen 2007). In the case of the logo, however, a collaborative wiki-solution was not possible. Not only is it technically demanding to change a 3-D image, but the logo itself had been registered as a trademark by the Wikimedia Foundation. This meant that any changes would require the official approval of the board. While some Wikipedians were content with the explanation that 'symbols were chosen because they looked good'[8] and that accuracy was not the main concern, others campaigned to have the errors corrected. In 2006, User:As, an IT professional from Bangalore in India, published a petition to change the misrepresented Devanagari characters. Nothing happened. In the same year, User:Mo, also from India, contacted Jimmy Wales directly, pointing out the mistake in the Devanagari script and asking for help to correct it.[9] Nothing happened. In 2007, another user contacted Jimmy Wales: he was informed that the errors were 'a known issue'.[10] Nothing happened. A year later User:Xe (from the United States) contacted User:No, the creator of the logo, asking for help to correct the mistakes. The answer was a monosyllabic 'no'. In the same year, User:Qu (from Canada) raised the issue on various forums, including Foundation-l. User:El, a board member at the time, promised to look into it, but also felt that it was not a serious issue: 'Not that the whole hubbub isn't a bit silly; the logo is meant to represent multilingualism as a concept, not any particular language.' In the end, nothing happened again, and it took seven years before the mistakes of the 2003 logo were addressed (Figure 4.5). The new logo, which was part of a larger design overhaul of the site by the Wikimedia Foundation, still appears not to be error-free and concerns have recently been raised about the representation of the Khmer script.[11]

Does the story of the Wikipedia logo matter? Or is it all just 'silly hubbub'? Digital multilingualism is not only about the number of webpages in different languages, the arrangement of languages on bilingual sites or the language choices made by bilingual users; it is also about how multilingualism is imagined on global sites such as Wikipedia, whose explicit aim and policy are to enable a more multilingual digital future.

The two 2003 Wikipedia logos expressed an overt and well-meaning commitment to multilingualism and diversity, and moved, symbolically, beyond the Anglocentricism of the early days. Yet the diversity symbolized by the logos was only skin deep. It was meaningful only to those who are not multilingual themselves, and would be satisfied with the aesthetics of 'cool looking' words and letters. Those who know different languages and scripts experienced especially the silverball logo by User:No as offensive. User:Ap, a senior contributor on the Japanese Wikipedia, comments on the Japanese reaction:

Figure 4.5 The Wikipedia logo: representing multilingualism III (May 2010)

> I know Japanese community is not happy with those two letters, and some takes it as a sign that the WMF has no interest towards Japanese projects and culture . . . its hardly to be considered a thoughtful design, rather a typical misrepresenting other cultures of Westerners (personally I love this kind, since it makes me . . . smile at least). (June 15, 2007; posted on Foundation-l)

Jane Hill's work (2008) on mock language provides a theoretical lens for thinking about the logo.[12] 'Mocking' refers to the action of imitating someone imperfectly, and by doing so expressing a derogatory attitude toward that which is being imitated. Usually mocking assumes intention and awareness. But sometimes our actions might seem mocking to others, even though we have no intention to mock. That is, the connotations of what we say or do go beyond what we consciously 'mean' or 'intend'. Hill develops her argument by looking at the use of Spanish words and phrases by non-Latino speakers of American English. Think of expressions such as *no problemo, mucho* or *hasta la vista*, which are used in a jocular way by non-Latinos, projecting a fun, easy-going and cosmopolitan persona. An important aspect of this linguistic practice is its disregard for the linguistic conventions of Spanish: bold mispronunciations and ungrammatical constructions are common. For example, Spanish speakers would not use the stereotypical /o/ pronunciation – *problemo, mucho* – but would say *problema* and *muy*. Speakers of Spanish perceive such language use – irrespective of its humorous intention – as offensive and disrespectful, as mocking their language. Hill draws on Eleanor Ochs' (1990) work on direct and indirect indexicality to describe how signs can hold two opposing meanings. Direct indexicality refers to what we believe we are doing, to our intentions. Thus, the intention of using Spanish words is to project a friendly and fun persona; the intention of the logo was to move beyond earlier monolingual representations and to showcase Wikipedia's multilingual knowledge project. Indirect indexicality refers to the unintended meanings that are available for interpretation. A long history of disregard toward non-European languages, and the Foundation's tardiness in addressing the issue, meant that the mistakes in the logo were read not simply as *mistakes*, but as a *sign* of lack of interest in languages other than English. (I will return to the question of intentionality in Chapter 5.)

Relationships of dominance and power are central to Hill's concept of mock language. This allows us to distinguish mock language from other practices, including texts that Jan Blommaert (2008) characterized as *grassroots*, that is, non-elite forms of writing, which were discussed briefly in Chapter 2. Both mock and grassroots practices produce texts with *mistakes* and *errors*, that is, texts that are experienced as *disorderly* because they violate existing norms of usage. Yet what distinguishes mock and grassroots texts is that they are produced from different positions of power. Mock texts are produced by empowered individuals, who reshape 'the meaning of the borrowed material into forms that advance their own interests', and who are able to impose their texts on others without being sanctioned (Hill 2008: 158; see also Bourdieu 1991 on 'strategies of condescension'). This applies to Wikipedia, where the majority of contributors are White and male and reside in the global North.[13] Their demographic profile is typical of those who are most empowered in the current world system, and, in the case of the logo, they were able to impose their design – mistakes notwithstanding – on others. Grassroots texts, on the other hand, are produced by writers who stand outside of global regimes of knowledge and are unable to impose their texts on others. This reflects their lack of power and leads to the limited mobility of texts (as discussed in Chapter 2). In other words, whether a text is mock or grassroots is determined not by the *intention* of the writer, or by the *attitude* of the reader, or by the *form* of the text (orderly or disorderly), but by the relative position of the writer/reader in the world system.

Wikipedians have worked hard to build Wikipedia's reputation as a source of knowledge that, in its English edition, equals that of established reference works such as the *Encyclopedia Brittanica* (Giles 2005). However, we know much less about Wikipedia's other language editions. The next section looks at Wikipedia's multilingual project, with particular focus on the isiXhosa Wikipedia as an example of a small and marginalized edition.

PRACTICES AND IDEOLOGIES II: SUPPORTING MULTILINGUALISM

New language editions are centrally administered by Wikipedia's Language Committee. In order to request a new language edition, one first needs to ensure that the language in question carries a valid ISO code, that is, has been assigned 'language status' by the International Organization for Standardization (ISO). Thus, similar to the way in which countries or currencies are identified by standardized codes, languages too have been assigned unique identifiers for technical or practical purposes (these codes are used, for example, when organizing books in libraries, or when documenting software localization projects). Since 2007, ISO codes for languages have been administered by the Summer Institute of Linguistics (SIL International). SIL International is a faith-based organization. It has close ties to Wycliffe Bible Translators, which is supportive of a project not unlike that of Wikipedia, that is, to provide every person not with an encyclopedia, but with the Bible in their own language. The very practice of assigning language codes is obviously problematic. It raises questions of legitimacy (who may assign such codes and on the basis of what principles?), as well as more fundamental epistemological questions of whether languages can be 'scientifically' classified (see Makoni and Pennycook 2007).

An ISO code alone does not guarantee that permission is granted to start a new Wikipedia edition. The applicant also needs to show that the language is 'sufficiently unique' and cannot be covered by an existing edition, and that there are enough 'living native speakers to form a viable community and audience'.[14] For those with limited online access, the practicalities of proposing a new language edition could be daunting. Obtaining codes and writing proposals is a challenge for mobile access, as discussed in the previous chapter. It is not impossible, but certainly not easy.

At the time of writing, Wikipedia was available in close to three hundred language editions and another three hundred languages were 'on trial', covering about 10 percent of the world's estimated six thousand or seven thousand languages. Wikipedia keeps track of its own multilingualism by producing extensive statistics, counting the number of different language editions, the number of articles in each edition, the most active editors and the number of edits they have made, and so forth. At this stage, Wikipedia's multilingualism is primarily a European affair. Although European languages[15] account for only a small proportion of languages spoken in the world, they dominate on Wikipedia in terms of editions as well as article count. Languages of Africa, America and the Pacific are severely underrepresented, and existing editions tend to be small. Asian languages do fairly well in terms of representation, but the majority of editions are small as well (Figure 4.6).

One of the smallest language editions is the one for isiXhosa, established in 2006, with a mere 203 articles in January 2014 (only Cree and Chichewa had less). The following discussion looks at the contributors to the isiXhosa Wikipedia and the texts they produce.

The majority of contributors to the isiXhosa Wikipedia come from the global North, the United States, Europe and Australia. A look at their user pages shows that they are not members of the diaspora, and most of them seem to be linguistic jacks-of-all-trades. There is, for example, User:Jo, a bilingual speaker of English and Chinese, and a veritable language enthusiast. In addition to isiXhosa, a language that – by his own admission – he cannot speak 'at all', he has been active on a number of other Wikipedias: Kikuyu, chiShona, Sesotho, siSwati, Kikongo, isiZulu, Luganda, Xitsonga, Somali, Oromoo, Chichewa, Sango, chiVenda, Northern Sotho, Ewe, Kirundi and Akan as well as Aromanian, Uyghur and Cherokee. There

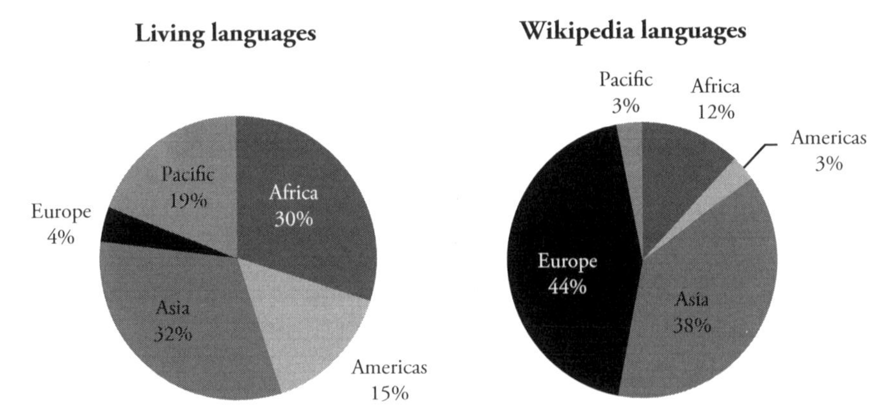

Figure 4.6 Spoken-language diversity and Wikipedia's multilingual landscape (language editions) (Lewis et al. 2013; http://stats.wikimedia.org/EN/Sitemap.htm, January 15, 2014)

is also the user from Phoenix, Arizona, who created two dozen articles for the isiXhosa Wikipedia in 2007, while also contributing to Maori, Min Nam, Zhuang, Cherokee, Limburgish and Sardinian. Many contributors from South Africa are in a position similar to that of those from abroad. They are English- or Afrikaans-speaking and have no or very limited competence in isiXhosa. Here too, lack of knowledge is matched by passion and enthusiasm. Consider the following statement on User:Ad's page:

(1) Molo! Unjani? Ndiright. Ndisasifunda isiXhosa. ['Hello! How are you? I am OK. I am still learning isiXhosa.'] Hi, my name is Adriaan. I am an Afrikaans speaking youth . . . I think it's absolutely sickening how the black languages are being marginilized in our country. I will try to help build the foundations of this Wikipedia. I have no formal schooling of Xhosa. The school taught us some Xhosa from grade 4 to grade 6, but it was mixed up with Zulu, and the teacher didn't even speak Xhosa correctly. So I will try to pick up some Xhosa along the way. I believe it's absolutely possible to learn Xhosa by myself.

User: Ad's post brings to mind well-documented colonial ideologies about African languages; namely, that they can be acquired easily and quickly by those who set their mind to it, and that second-language speakers can make meaningful contributions to the language's development (Comaroff and Comaroff 1991; Schieffelin 2000; also Greenblatt 1991). The practice of editing other-language Wikipedias is often framed as a desire 'to help' by Wikipedians, that is, to make marginalized languages visible and to contribute to Wikipedia's multilingual vision. Such discourses of benevolence – rhetorical strategies that encourage people 'to do well' toward those considered to be less fortunate – are closely linked to what Monica Heller and Alexandre Duchêne (2007) have referred to as 'discourses of endangerment'. The latter reflect popular and academic concerns about the worldwide loss of linguistic diversity, and the belief that community outsiders can play a role in turning the tide (see also Perley 2012 for a critical discussion). Such sentiments are found not only among the rank and file, the ordinary contributors, but also among senior Wikipedians, such as User:Mi.

> (2) It is likely that all of 3000 languages with more than 10,000 speakers would survive if they have Wikipedia edition in their language. And if you ask why Wikimedia movement should do that, it's because there is no other relevant international body capable to do that. That makes Wikimedia's position unique and with large amount of historical responsibility. (http://wikimedia.7.x6.nabble.com/Wikimedia-l-Thanks-for-all-the-fish-td5006698.html; June 2013)

Wikipedians support the project's multilingual vision in various ways. Some engage in cleaning-up edits on other-language Wikipedias – that is, edits that remove nonsense or offensive material created by digital vandals – or they simply work on the design and layout of the page. However, many seem to want to do more and create, their lack of proficiency notwithstanding, entire articles from scratch. But how do you contribute content if you don't speak the language?

User:Ka did not contribute to the isiXhosa Wikipedia, but to the Otjiherero Wikipedia. He describes how he went about publishing articles in Otjiherero, a language he did not know (Otjiherero is spoken in Namibia and Botswana by just over two hundred thousand speakers).

> (3) I heard about the Herero language some years ago, and since February (I think) i discover the wikipedia in this language; sadly at that time the Wiki had less than 5 articles, and with a little searching I found articles about religion in that language. at first I deduced comparing the text with the spanish version, and I extracted some related religios (*ozomganburiro*) words. Now I'm triying to deduce other words related to life (*Omuinyo*) or maybe about earth (*Kombanda yehi*). All languange can be understand comparing with anothers, you can understand the composition of the words, the suffixes and prefixes . . . Sorry for my bad english:([16]

In this text, User:Ka outlines the basics of linguistic fieldwork as conducted by generations of missionaries, colonial officials and linguists. One starts by comparing sentences in a language one knows with parallel sentences in the target language. This allows one to understand the meanings of words and to infer the rules of grammar. And finally, one starts producing texts. Others might work less empirically and draw more strongly on dictionaries and grammars (User:G, email communication, 2011). In the isiXhosa Wikipedia we can distinguish four main text types produced by linguistic and cultural outsiders: disorderly texts, pseudo-sentences, word lists and partial translations.

In 2005, an entry for 'sociology' was created on the isiXhosa Wikipedia by the editor from Phoenix, Arizona, mentioned earlier. The entry itself is incomplete (indicated by punctuation that signals ellipsis) and rudimentary, especially when compared to the equivalent English entry, which contains over 13,000 words (including extensive references). The text, accompanied by images of six well-known Northern sociologists (Comte, Durkheim, Pareto, Tönnies, Simmel and Weber), reads as follows:

(4) enzululwazi ngoluntwini benzakalisa esimnandi iifemeli . . . (xh.wikipedia, April 2014)

When I asked speakers of isiXhosa about this text, they were adamant that the text makes no sense at all and described it as 'just jumble'. One speaker, in his forties, commented as follows:

(5) Maybe they were trying to say *inzululwazi ngezoluntu* . . . which could be social scientist or sociologist, not sure . . . *enzekalisi ifemeli* means hurts the family, the rest doesnt make sense. (Email, December 2010)

This is echoed by a second speaker, a South African student in his early twenties:

(6) *Benzakalisa* means hurting someone or something; however, it does not make sense! *Esimnandi* meaning something nice – more like something you enjoy especially, well in most cases referring to food! NB: please try to make sense of the meaning and explanation of these words from this text of sociology because i cant figure it out what they are trying to say!! (Email, December 2010)

Other articles are problematic in different ways. So-called city-spam entries are a prominent feature of smaller Wikipedia editions. City-spam entries allow contributors to create pseudo-sentences based on structural repetition, similar to drill exercises in the second-language classroom (Schieffelin 2000). On the isiXhosa Wikipedia, for example, an editor from Bad Homburg in Germany introduced the phrasal pattern 'X is a town', *idolophu* or 'village', *ilali*, 'in Y' (name of country). The phrase was picked up by editors from across the world to create structurally identical sentences, which are grammatically well-formed, but are unlikely to interest potential readers.

(7) *iHanoveri idolophu eJamani* (created by a German user in 2007)
Varenholz ilali eJamani (created by a German user in 2007)
Hamburg idolophu eJamani (created by a Dutch user in 2008)
Pigazzano idolophu eItaly (created by an Italian user in 2009)
Osaco idolophu eBrazil (created by a Brazilian user in 2011)

Other pages reflect a dictionary-like preoccupation with words and word lists. The format is always the same: a single word in the target language is accompanied by a photograph or image that illustrates its meaning, Figure 4.7 shows the entry for *ikofi*, 'coffee'. Other examples include *inja* accompanied by the image of a dog, *ixoxo* and the image of a frog, *intaba* and the image of mountain, *incwadi* and the image of books, and so forth.

And finally, there are partial or incomplete translations. Thus, in early 2014,

Figure 4.7 *Ikofi* is 'coffee', created in March 2014 by User:Vi, an 'English speaking American from Chicago' (May 14, 2014)

the interface of the isiXhosa Wikipedia remained a mixed affair. Some parts were translated, others were still in English, awaiting someone to come along and step in. Not only links but articles too can be partially translated and combine different languages. For example, the start-page of the isiXhosa Wikipedia was, at the time of writing (April 2014), still a bilingual, partially translated text.

Speakers of isiXhosa have on the whole been silent, neither expressing offence, nor correcting, wiki-style, what is there. The texts currently on display – mistakes notwithstanding – are unusually stable, and the entry on sociology has not been corrected or commented on in almost ten years. The texts on the isiXhosa Wikipedia create a peculiar linguistic landscape. They display words and grammatical structures that can be identified as 'isiXhosa', yet at the same time they fail to represent 'isiXhosa'. They are reminiscent of Jean Baudrillard's (1983) notion of the simulacrum, an image that bears only an indirect relation to reality. The simulacrum evokes and approximates; it creates a new reality rather than merely reflecting an existing one (see also Hall 1997). And in doing this the texts on the isiXhosa Wikipedia reproduce relations of power and inequality, reinscribe existing hierarchies and affirm the marginal status of isiXhosa online.

PRACTICES AND IDEOLOGIES III: BETWEEN BOTS AND SUBVERSION

Like everything online, Wikipedia is a moving target, and the articles it contains are necessarily 'unfinished artefacts in a continuing process' (Bruns 2008: 110). Change may have been slow so far, but is always possible, and what I have described above can change quite quickly. In early 2012, Douglas Scott, an English-speaking South African Wikipedian, organized a series of workshops in Cape Town to encourage first-language speakers to contribute to the isiXhosa Wikipedia. He targeted African language departments at local universities and introduced academics, as well as students and language professionals, to the practicalities of editing Wikipedia. There was an immediate benefit in terms of growth, since at the workshops each participant would create at least one new article. As a result, many of the articles created in 2012 were produced by isiXhosa speakers. The new articles were substantial in terms of length and appropriate in terms of language use. Further entries were composed in August 2013 by students from Sinenjongo High in Cape Town as part of a competition. In March 2014, as I was editing this chapter, an isiXhosa-speaking contributor, User:No, appeared. He created several new entries, edited existing ones and even provided impromptu language lessons for those wanting to contribute. Whether any of these activities will lead to more sustained, *collective* engagement by first-language speakers remains to be seen, and, at the time of writing, language enthusiasts from the global North were still a feature of the isiXhosa Wikipedia.

While most African-language Wikipedias continue to languish, some show substantial activity. In early 2014, the top African Wikipedias were Malagasy, Yorùbá, Afrikaans and Kiswahili, with Egyptian Arabic catching up fast. In the case of the latter three, there exists a core group of editors – either first-language or high-proficiency second-language speakers – who keep the project going collectively. Things are different on the Yorùbá and the Malagasy Wikipedias. In early 2014, the Yorùbá edition had over 30,000 articles, the Malagasy edition almost 50,000. Each of these Wikipedias was created almost single-handedly by just one person: User:Ja, who belongs to the Malagasy diaspora in France, and User:De, a Yorùbá speaker. Such individual dedication is not unusual for Wikipedia; even on the English edition 'a small proportion of editors account for most of the work' (Panciera et al. 2009). However, the remarkable growth of these two editions was not the result of a lone writer spending hours in front of his computer, composing and translating articles. Rather, articles were created with the help of bots, that is, software applications that run automated translation tasks. This allows for the speedy creation of thousands of short, stub-like articles and increases the article count (as a measure of 'success').

Bot-created articles, which have appeared on Wikipedia since 2002, typically provide limited information to the reader and can be of dubious quality (depending on the translation software). Consider, for example, the entry for 'internet' on the Yorùbá Wikipedia given in (8). There are similarities to the 'sociology' entry discussed above: both texts are short and contain mistakes. However, the errors in the Yorùbá text – missing and misplaced diacritics and some odd phrasing due to direct translation from English – are less fundamental than those in the 'sociology' entry, and the text is comprehensible to speakers ('awkward yet one can make sense of it'; Tolulope Odebunmi, personal communication, 2013).

(8) Internet jé asopò bí ìtakùn àwon erò kòmpútà kakakiri àgbáyé fún ìpàsíparó ìpolongo. Ní èdè Gẹ̀ẹ́sì ó wá láti ìsopò *inter-network*. (April 2014; page was created in 2008 by User:De)

> *Translation:* 'The internet is like a web and links computers from all over the world with the purpose of exchanging information. The English word is a combination of *inter-network*.'[17]

The use of bots is not limited to African-language Wikipedias, and two of the top fifteen Wikipedias in terms of article count – Cebuano and Waray-Waray (both spoken in the Philippines) – contain over 80 percent bot-created material. Bots have also been important in the development of at least two major European Wikipedias: the Dutch and the Swedish. András Kornei (2013) refers to such practices as 'gaming the system' and describes the resulting editions as Potemkin Wikipedias. They are created to impress, yet there is ultimately little substance to them. One is reminded again of Baudrillard's notion of the simulacrum. Bots create a proliferation of representations that have no real origin, no ground or foundation. The landscape they create is in some ways similar to as well as different from that created by the human editors on the isiXhosa Wikipedia. Repetitive, stub-like entries abound, but texts are, on the whole, comprehensible, even though sometimes awkward and unidiomatic.

Technologically enabled multilingualism is not limited to Wikipedia, and tools such as Google Translate allow users to produce approximate texts in a variety of languages. The use of Google-assisted multilingualism is also an increasingly common practice on social network(ing) sites, where 'users suddenly come up with phrases in a language that . . . they have no command of. For example, speakers of Greek or English may come up with a few Mandarin Chinese signs, non-speakers of German with a slightly unidiomatic or ungrammatical phrase in German, and so on' (Androutsopoulos 2013: 5; see also Barton and Lee 2013: 63).

While some engage with the Wikipedia project on its own terms, others challenge the very foundations of Wikipedia's knowledge project. On Wikipedia, *notability* and *verifiability* are defining principles and shape representation. Articles should only be created about topics deemed 'notable'. But how do we know that something is notable? Here verifiability comes in as a way to ensure the quality of entries. If something is of general rather than personal interest, then others will have written about it and it is possible to cite 'reliable', 'published' and 'independent' sources; that is, the information can be verified according to conventions familiar from academic writing.[18] However, unlike in academic writing, original research is not permissible on Wikipedia, and this raises questions about the global politics of knowledge: what about topics that have not yet attracted the attentions of scholars and journalists? Can we not write about them? And what about oral sources? Are they allowed on Wikipedia? (See the articles in Lovink and Tkacz 2011.)

During the 2012 isiXhosa Wikipedia workshops, one participant quietly subverted the instructions given. He did not translate existing articles from English into isiXhosa (as was suggested), nor did he write new articles based on Wikipedia's principles. Instead, he created an isiXhosa page that turned Wikipedia's rules on their head: *iziduko zethu* 'our clan names'. AmaXhosa clan names reflect shared descent, history and kinship, and every clan name has a set of so-called clan praises

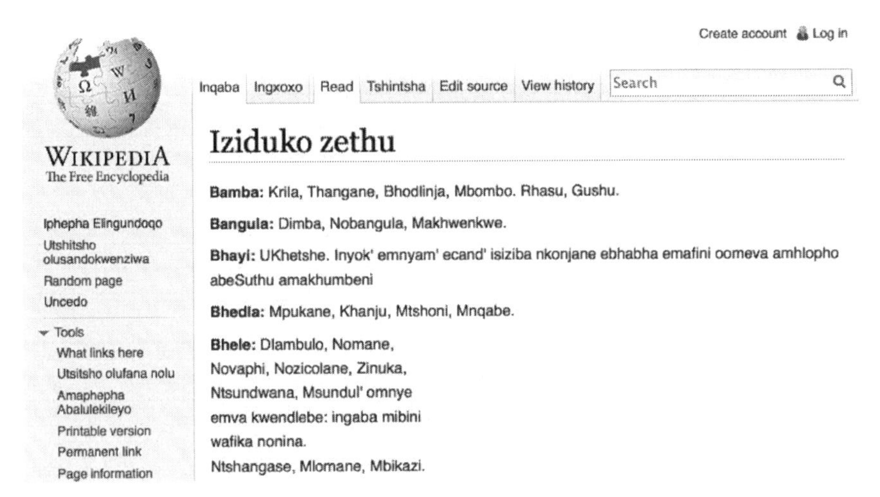

Figure 4.8 *Iziduko zethu* 'our clan names' by User:Ab, an isiXhosa-speaking South African (April 2014)

attached to it. These praises trace the lineage of the clan sequentially, from the most distant (male) ancestor to the most recent one. The author of the page lists – without referencing any written sources or providing explanatory prose – the praises for the Bamba, Bangula, Bhayi, Bhedla and Bhele; praises for the Dlamini, Miya and Togu were added later and the entries for the former were expanded (Figure 4.8). The resulting page is locally meaningful and displays culturally embedded language, but is unlikely to be understood by cultural outsiders.[19]

The page violates the editorial policies of Wikipedia outlined above. Rather than providing an encyclopedic definition of the type 'Clan praises are . . . ', drawing on easily verifiable written sources, the author treats Wikipedia as a public archive and repository for the preservation of *oral knowledge*. The title of the entry is programmatic. This is not merely a page about 'clan names' or 'amaXhosa clan names', it is about '*our* clan names'. The personal rather than neutral tone is echoed on the associated talk page, where the writer addresses possible contributors to the page as *mawethu* 'my people', 'my kinfolk'. On the page we see 'real' and not 'simulated' isiXhosa. The clan praises are given as they would be recited in spoken interactions, and their representation is culturally intelligible to speakers of isiXhosa.

The issue of oral sources is not absent from Wikipedia's internal discussions, and in 2012 Achal Prabhala, a Bangalore-based member of Wikimedia's advisory board, conducted the Oral Citations Project – funded by the Wikimedia Foundation. His report stated programmatically: 'People are knowledge', and Prabhala created a number of articles about traditional Indian games based on his research on oral narrative and history.[20] While there is little policing of content on the isiXhosa Wikipedia, and oral sources slip through the cracks, the situation is different on the English version, which is heavily policed, and such articles – including the ones published by Prabhala – are quickly flagged as being in danger of deletion unless 'verifiable sources' are added. *Iziduko zethu* escaped such a fate and remains, for the time being, a feature of the isiXhosa Wikipedia.

CONCLUSION: NEW LINGUISTIC LANDSCAPES, OLD INEQUALITIES

This chapter has looked at the internet from the perspective of linguistic landscaping. Broad, global statistics show that the virtual space has diversified since the 1990s, and that English, although still strong, no longer dominates. However, sociolinguistic studies also show that only a few languages shape this diversity and that the majority of the world's languages are absent, or only minimally represented, online. An important question for digital multilingualism is thus: what will happen to these languages? That is, what will happen to the small languages, the minority languages, the languages with little power or 'currency' on global platforms?

Wikipedia's multilingual policy is an attempt to address this imbalance and to provide an inclusive space for multilingualism (or at least for those languages that have a valid ISO code). Yet Wikipedia is also an ambiguous space. The story of the logo – an integral part of the encyclopedia's semiotic landscape – showed disregard for the realities of lived multilingualism, and the uncorrected logo, displayed for seven years, displays features of what Hill calls mock language. The texts found in small Wikipedia editions, such as the isiXhosa one, are also problematic, but in a different way. They too are disorderly, that is, they diverge from existing norms and they contain mistakes. However, unlike the logo they are open to correction, and it is the absence of first-language speakers in these spaces that disrupts the logic of the wiki-system. The texts that are there create a linguistic landscape that, like the one created by bots, is not a representation of a real, existing language-in-use, but is better understood as a simulation. It is a sign for which there is no referent in the world, but which creates its own reality. It is 'a real without origin or reality' (Baudrillard 1983: 1), a *representational proxy* for the language we call 'isiXhosa'. These texts thus reproduce rather than challenge global inequalities. In other words, some languages are *represented* on Wikipedia and are displayed according to existing norms and conventions; other languages are *simulated* and exist only as approximate, disorderly or minimal texts.

A common assumption underlying much multilingual rhetoric is that people will necessarily want to access material in their first or 'native' language. However, in an increasingly global and interconnected world, where most people speak more than one language, such monoglot ideologies have little currency. This is particularly true for postcolonial societies where everyday multilingualism is the norm, and where the process of becoming literate is usually linked to acquiring proficiency in the former colonial language. Wikipedia page statistics show a clear preference, across Africa, Asia and South America, for reading – as well as contributing to – the large and comprehensive editions that are available in the former colonial languages. Jan

Blommaert (2010: 46f.) reminds us in his work on globalization that the symbolic upgrading of marginalized languages – promoted in the global North as a recognition and celebration of diversity – is often met with suspicion and resistance from those who are being 'helped' and 'supported' and who see such efforts – and the discourses of benevolence that underpin them – 'as an instrument preventing a way out of *real* marginalization and amounting to keeping people in their marginalized places and locked into one scale-level: the local'.

NOTES

1. http://www.bbc.co.uk/news/science-environment-17081573.
2. http://en.wikipedia.org/wiki/Wikipedia:The_rules_are_principles.
3. Pimienta and his colleagues (2009: 13) selected fifty-seven keywords for their analysis.
4. http://stats.wikimedia.org/EN/Sitemap.htm.
5. Reproductions of past Wikipedia logos can be found at http://en.wikipedia. org/wiki/Wikipedia:Wikipedia_logos.
6. Although Wikipedia data is in the public domain, I decided not to identify Wikipedians – except for Jimmy Wales – by their names/nicks, and have anonymized their contributions by using only the first two letters of their user names. I also do not provide direct links to named talk and user pages in order to protect contributors' privacy as much as possible (see Chapter 2 on ethics).
7. http://meta.wikimedia.org/wiki/Talk:Errors_in_the_Wikipedia_logo/Archive_1, and http://itre.cis.upenn.edu/~myl/languagelog/archives/004653.html.
8. http://meta.wikimedia.org/wiki/Talk:Main_Page/Archives/2006/01#Error_in_the_Wikipedia_Globe.
9. http://en.wikipedia.org/wiki/User_talk:Jimbo_Wales/Archive_16.
10. http://en.wikipedia.org/wiki/User_talk:Jimbo_Wales/Archive_25.
11. http://meta.wikimedia.org/wiki/Talk:Wikipedia/Logo#Khmer_font.
12. Although I am using Hill's terminology here, there are others who have worked on similar phenomena but referred to them differently, for example, 'astroturf literacy' (Vigouroux 2011) or 'fake multilingualism' (Kelly-Holmes 2005).
13. http://en.wikipedia.org/wiki/Wikipedia:Systemic_bias.
14. http://meta.wikimedia.org/wiki/Meta:Language_proposal_policy.
15. English, French, Portuguese and Spanish are counted as European languages because of language typology and history.
16. http://meta.wikimedia.org/wiki/Proposals_for_closing_projects/Closure_of_Herero_Wikipedia (2007).
17. A big thank you to Tolulope Odebunmi for her patient help with the Yorùbá data.
18. http://en.wikipedia.org/wiki/Wikipedia:Notability. Stylistically too, Wikipedia emulates traditional academic prose, and articles should be written from a 'neutral point of view' (NPOV); that is, writers should not take sides and should use a maximally objective voice.
19. User:Gu (from the Netherlands) mentioned similar experiences when I asked him about his involvement with the Bambara Wikipedia. During a stay with the Geekcorps in Mali, he paid Bambara speakers to write short articles: 'They had written about topics like milk, a specific tree, their local village, which are all great for Wikipedia. Some of them didn't get the initial "assignment" in the way I meant it and wrote poetry (I still paid him for that and put it somewhere online . . . not sure where anymore)' (email, March 2011).
20. http://meta.wikimedia.org/w/index.php?title=Research:Oral_Citations&oldid=3171583.

Chapter 5

Intertextuality and author-audiences

> I can't help it; I am a bit of a jukebox of intertextuality. If someone says a phrase; I continue it with a lyric I know.
>
> Twitter 2013

INTRODUCTION: CIRCULATING TEXTS

Virtual linguistic landscapes, as discussed and illustrated in the previous chapter, are multimodal, and combine images and linguistic material in various ways. This chapter continues the discussion of multimodality, focusing on user-generated content on YouTube. Following Jannis Androutsopoulos (2010) and other media scholars, YouTube videos are conceptualized as spectacles, that is, public performances that emphasize the visual, and are intended to be seen, not just listened to or read.

A central theoretical concept in the discussion is the notion of intertextuality. The term was introduced by the French philosopher Julia Kristeva ([1969] 1980) in the 1960s, and is now a core concept in media theory as well as sociolinguistics and linguistic anthropology (see Allen 2011). Intertextuality describes the way in which texts don't stand alone, but are connected to one another, and carry traces of other texts, in both form and content. It is important to note that text is used here somewhat differently from the way it is used in everyday speech. Conventionally when we talk about a text we are thinking about a piece of writing, such as Leo Tolstoy's *Anna Karenina*, or the text of a speech to be delivered. Yet texts – in the sense the word is used here – are not tied to writing; rather, texts are understood broadly as anything spoken, written, filmed, danced, painted or sung *that can be interpreted by an audience.*

As audiences engage with, and respond to, what they read, hear or see, they create new texts that are equally embedded in networks of meaning. The read/write environment of Web 2.0 technologies provides important opportunities for audience engagement. Web 2.0 audiences express their appreciation – or lack thereof – through *like* buttons (an online version of applause) and written comments as well as amateur spectacles that creatively rework existing materials. Such activities recontextualize existing material, often localize it, and offer interpretations that can be quite different from the originating text.[1]

Although the focus of the discussion in this chapter is the digital realm, audience engagement is not limited to online environments, and there exists a complex interplay of offline and online worlds. Consider 'Kony2012', a short film that was published by Jason Russell on YouTube (2012). The aim of the video was to promote a campaign to stop Joseph Kony, a Ugandan warlord, and to have him arrested before the end of the year. The video spread quickly and many supported the campaign by talking and tweeting about it, raising awareness on Facebook and donating money to Invisible Children, the charity behind the video. But not all responses were favorable. Soon questions were raised about the integrity and honesty of Invisible Children; the video itself was criticized as being simplistic, sensationalist and inaccurate, and those supporting the campaign as politically uninformed and foolish. As criticism of the campaign grew and became more vocal, Invisible Children closed the comment function on YouTube. Yet debates continued online *and* offline (Waldorf 2012). Figure 5.1 comes from a female toilet cubicle at the University of Cape Town. The writer asks those who enthusiastically supported the campaign: 'Do you even know where Uganda is?', and receives several iconic 'likes' from readers in response.

Figure 5.1 'Do you even know where Uganda is?' Offline audience engagement with a YouTube event ('Kony2012') reframes digital practices (© Carolyn Le Tang, 2012)

DIGITAL SPECTACLES: MULTIMODALITY AT ONE'S FINGERTIPS

Communication is never about language alone. Gestures, gazes, movements, bodies and their adornments, as well as material contexts – such as meeting someone in a cozy bar as opposed to a busy supermarket – shape meaning and interaction. In digital communication too different modalities can be combined: image, color, layout and text as well as sound and video. New media thus provide us not only with interactional affordances (Chapter 2), but also with multimodal affordances (Kress 2010). These create a semiotically rich environment that, as discussed in Chapter 2, is different from body-to-body interaction, but not impoverished in its expressivity.

In 2010, for example, the multimodal affordances of digital communication allowed Chinese dissidents to escape political surveillance and to play cat-and-mouse with government censors. In that year, Liu Xiaobo, who was serving a jail sentence for political activism at the time, was awarded the Nobel Peace Prize. The Chinese government, known for persistent online censorship, initially blocked all digital messages containing the name of the dissident laureate. Soon they blocked further *words* in order to silence unwanted political debate, including 'Norway', 'Oslo', 'Nobel' and even 'empty chair', the latter the most conspicuous symbol of Liu Xiaobo's absence at the ceremony in Oslo. Chinese bloggers responded to these *linguistic* restrictions creatively by posting *images* of empty chairs, ranging from reproductions of a Van Gogh painting to seemingly arbitrary photographs of kitchen chairs. By exploiting the multimodal affordances of the medium, the bloggers kept the discussion alive and avoided the censors just a little longer (Larmer 2011; see also Link and Xiao Qiang 2013).

The multimodal displays typical of Web 2.0 environments have been described as spectacles by media scholars; that is, as public performances that are meant to entertain, impress, shock, provoke or amuse mass audiences (Best and Kellner 1999; Darley 2000; Kellner 2009; Androutsopoulos 2010). The noun 'spectacle' derives from the Latin root *spectare* 'to look at, to watch'. The visual aspect is essential, and spectacles hold our gaze and engage our eyes. Moreover, spectacles are, by definition, spectacular events. Their displays are attractive, dramatic, exuberant and dazzling, and provide us with an experience set apart from everyday life (Sutherland 2012). Like so much in the social world, spectacularity is a matter of degree. Spectacles range from global mega-events, such as the Olympics or the Soccer World Cup, to national election campaigns, local music festivals and amateur YouTube videos. We can even make 'a spectacle of ourselves' when we behave in ways that are considered excessive. Increasingly spectacles are mediatized, and frequently the 'real' event is orchestrated in such a way that it optimally displays on computer and television screens. The Olympics, for example, are today essentially an event made for television (and increasingly made for the internet). Only very few people watch the games live, and offering a telegenic experience is essential for audience success, that is, the popularity, and thus marketability and dollar-value, of the games (Billings 2008).

Critical theorists, working from a broadly Marxist perspective, have treated mass-mediated spectacles with suspicion. A landmark text was Guy Debord's manifesto *The Society of the Spectacle* ([1967] 1994). Debord argues that spectacles are hegemonic and authoritative events that represent 'the existing order's uninterrupted discourse about itself, its laudatory monologue' (Thesis 24). Debord sees these monologic spectacles as a tool for social pacification and depoliticization. Spectacles, according to Debord, build an 'empire of modern passivity' (Thesis 14), and alienate us from our own desires, thoughts and feelings (Thesis 30). For Debord and other critical theorists (e.g. Horkheimer and Adorno [1944] 2002; Marcuse 1964), spectacles entertain, amuse and manipulate us through 'shimmering diversions' (*Thesis 59*), ultimately leading to social isolation, apathy, anomie and alienation. The image that comes to mind is that of a lone individual sitting in front of a television or computer screen. Yet these views did not go unchallenged, and others have emphasized the agency and creativity of spectators. Rather than passively consuming the images and sounds before their eyes in couch-potato style, audiences respond to, engage with, critique, contest and appropriate the displays of the spectacle: audiences applaud, boo, comment, assess and evaluate; they re-enact and parody what they see and hear. In other words, as they are watching they are engaging with the – indeed often authoritative and monologic – performance on display as well as with one another (see, e.g., Beeman 1993; Manning 1996; Stevenson 2010).

In *Forms of Talk* and *Frame Analysis*, Erving Goffman (1981, 1974) distinguishes between game, or 'inner realm', and spectacle as two distinct, but deeply interrelated, interactional frames, that is, conventions through which we organize and interpret – frame – our experiences. In Goffman's terminology, game refers to the events that are on display, such as a soccer match or a political rally. In his later work (1983), he calls such displays 'platform events' (1983). Spectacle, according to Goffman, refers to the broader context: the activities, sideshows and off-stage interactions, that is, the *mise-en-scène* that surrounds the center-piece (on-stage) performance, and that turns the event into a collectively experienced 'social occasion' (Figure 5.2).

Games or platform events are, according to Goffman (1981, 1983), focused gatherings. They are spatially and temporally bound, participants have well-defined

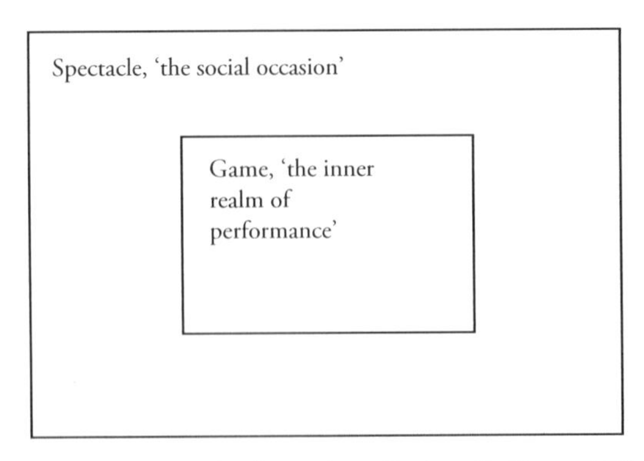

Figure 5.2 Game and spectacle (based on Erving Goffman 1981, 1974)

roles, and their behavior is guided by scripts and rules. As such, games differ qualitatively from the more diffuse interactions and happenings of the larger spectacle.

It is analytically useful to separate focused from diffuse gatherings, planned and scripted events from their unscripted, spontaneous and sometimes anarchic encasings. However, Goffman's distinction between game/platform event and spectacle/social occasion is not commonly used in the literature, and most writers employ the term spectacle to describe the entire event. That is, a spectacle is seen as a multi-authored mega-genre – involving both scripted on-stage events and diffuse backstage events – in which 'audience and performer share a joint expressive burden' (Manning 1996: 264). While live spectacles require the co-presence of performers and audiences, mass-mediated spectacles can reach audiences in different locales, and re-screenings allow audiences to engage with the performance months or even years later.

New media, and especially Web 2.0 environments, have multiplied and transformed opportunities for audience participation. Potentially everyone is now a producer, and visual spectacles – albeit on a significantly smaller scale than a Hollywood blockbuster or a TV show – can be designed and distributed using one's computer or even a high-end mobile phone. Jannis Androutsopoulos (2010) describes Web 2.0 platforms such as YouTube (founded in 2005) as enabling complex participatory spectacles. The center-piece performances are video clips that are uploaded and can be viewed by a potentially global mass audience. Audience participation is possible through likes and dislikes – represented visually as 👍 and 👎 – as well as via the comment function, which allows viewers to post their opinions about the video, or to respond to the views of others. A more time-consuming option is to produce response videos. These can be in the form of a parody, but can also celebrate aspects of the original. Figure 5.3 illustrates this encasing of activities – the 'laminated affair of spectacle and game' (Goffman 1981: 167) and the interconnected meaning-making – that is, intertextuality – that defines spectacles.

The example is Barack Obama's *Yes we can* speech, which formed part of the truly spectacular race for the US presidential election in 2009 (Kellner 2009). The speech, delivered to a live audience in 2008, was uploaded a day later on YouTube by BarackObamadotcom, Obama's official account. The video of Obama's speech is surrounded by a diversity of audience responses. This includes likes and dislikes, comments and response videos. I will discuss this particular example in more detail later in this chapter. While Figure 5.3 concentrates on just one application – YouTube – audience responses can occur cross-platform. People wrote about the speech on blogs, tweeted about it and discussed it on Facebook. And, as in the case of 'Kony2012', we see a complex overlap of online and offline engagement. People also talked about it body-to-body and on radio shows, wrote about it in newspapers, and so forth. This creates chains of communication on multiple scales, from local to global, from professionally produced to privately created, from individual to collective, from old to new media (Agha 2011; Blommaert 2010: 32ff.).

Figure 5.3 Barack Obama's *Yes we can* speech as a YouTube spectacle: frames within frames (based on YouTube data collected February 2013)

INTERTEXTUALITY: THEORETICAL REFLECTIONS

The different frames in Figure 5.3 – from the center-piece to likes/dislikes, comments and response videos – are rich in intertextuality. They orient toward each other, as well as toward earlier texts. In new media studies the terms mash-up and remix cover a conceptual terrain that is similar to that of intertextuality (Jones and Hafner 2012: 45ff.). However, whereas intertextuality describes a fundamental characteristic of all signs, mash-up and remix refer more narrowly to the technical processes through which digital artifacts are created, especially the medium-specific ease with which materials can be copied, combined and edited. This section provides a general theoretical discussion of intertextuality, bringing together contributions from linguistic anthropology (Bauman and Briggs) and language philosophy (Derrida).

Richard Bauman and Charles Briggs (1990) provide us with a model of how intertextuality comes about. Central to their argument is the terminological trio of entextualization, decontextualization and recontextualization. The core concepts are text and con-text. While *text*, as noted above, refers to any combination of signs that is interpretable by others, *context* describes the broader environment (social, psychological and linguistic) in which a text occurs (and that shapes its interpretation; Hanks 1989). Entextualization is foundational to intertextuality. It describes the general process by which a stretch of continuous semiotic production is perceived as a self-contained, bounded object; that is, as a text, which can be copied or imitated, lifted out of its 'infinitely rich, exquisitely detailed context', and reinserted into new, equally rich but different contexts (Silverstein and Urban 1996: 1). Within this framework the notion of a text takes on an additional meaning. It is not only *interpretable*, but also *detachable*. Entextualization often proceeds along generic lines. That is, culturally known genres, such as stories, songs and jokes, are experienced as internally cohesive, and are retold and resung with ease. However, it is also possible for only part of a semiotic sequence to be reproduced. In Chapter 2, I mentioned the Korean music video 'Gangnam Style', the YouTube sensation of 2012. The four-minute video shows the South Korean musician Psy performing the song while moving through a range of scenerios, from horse stables to playgrounds and parking garages. The dance routines that accompany the music are smooth, integrating a wide range of – sometimes intentionally comical – dance moves. One particular set of moves stood out in the reception of the video: the invisible-horse-riding dance. The dance became a central feature in various response videos, was dubbed a 'craze' in global media commentaries and was performed in offline contexts. For example, around the time of Psy's YouTube success it was not unusual to see teenagers and children in the streets of Cape Town riding their invisible horses. The process of entextualization was supported by the fact that the dance occurred repeatedly throughout the video. It was performed by Psy alone as well as by a crew of dancers. Thus, out of the continuous dance performance in the video, a particular move emerged – through repetition – as a bounded routine, a collection of dance steps that could be copied and reproduced on a global scale.

Once a perceivable and reproducible text has emerged through entextualization, it becomes available for the potentially infinite processes of decontextualization and recontextualization. The text is detached from the context in which it occurred and recontextualized, that is, inserted into new contexts.[2] Thus, decontextualization and recontextualization are not separate processes, but always occur in tandem:

'decontextualization from one context *must* involve recontextualization in another' (Bauman 2004: 4; my emphasis). When inserted into new contexts the replicated text might look the same, but it will not mean the same, and although it carries with it certain meanings from its earlier uses, it also acquires new meanings. For example, as a narrative or song is lifted out of orality and is recontextualized in writing the change in medium brings with it a change in meaning, as writing often imbues texts with a sense of permanence and authority. There might also be a change in genre (e.g. when a narrative or song is reproduced in a scholarly article to illustrate a particular point), as well as audience (the text can now be distributed beyond the local conditions of its production).

Intertextuality is not something that happens only in literature or new media environments, but is fundamental to the way language works. Speaking always and necessarily involves the use of words, phrases and expressions that existed prior to the here and now, and as we use them and 'push them into new contexts', we inevitably orient toward their earlier uses (Becker 1995: 185). An understanding of language as fundamentally intertextual – that is, as based on ongoing and repeated processes of recontextualization – has been present in linguistic theory for a long time. Thus, when Humboldt (as quoted in Chapter 2) notes that the foundation of speaking (and writing) 'lies in the speaking and having spoken of all previous generations', he articulates a view of language as essentially intertextual. The Russian language philosopher Mikhail Bakhtin, whose work will be discussed in greater detail in the next chapter, expressed a similar perspective roughly one hundred years later when he wrote: 'The word in language is half someone else's . . . it exists in other people's mouths, in other people's contexts, serving other people's intentions: it is from there that one must take the word and make it one's own' ([1934/1935] 1981: 293–4). Repetition – far from being boring and unimaginative – is thus central to the creativity. As we draw on 'other people's words' and repeat them, we also reshape them and make them work in new contexts (Toolan 2012; Tannen 2007). Creativity is located right there, at the interface of the old and the new, sameness and difference.[3]

Taking its cue from J.L. Austin's speech act theory, the Algerian-French philosopher Jacques Derrida ([1972] 1988) provides a spirited argument for the importance of repetition in his essay 'Signature Event Context'. Derrida has become more visible in sociolinguistic theory in recent years (Cameron and Kulick 2003; Pennycook 2007; Hodges 2011; Nakassis 2012), yet his work remains somewhat on the margins, possibly due to its often impenetrable nature. Let me try to unpack Derrida's response to Austin and his argument about the importance of repetition in language.

In *How to Do Things With Words* (1962), Austin discusses speech acts or performatives, that is, utterances that produce social actions and have the power to change the world around us. This is most clearly visible when one looks at explicit performatives such as apologies, blessings or promises that *do* things by saying them. (I will return to other types of performatives in the following chapter.) Thus, 'I do take this woman/man to be my lawfully wedded wife/husband' seals a marriage in a legally binding sense and produces a new reality for those involved. However, if the phrase is said by someone who is not sincere about it, who does not actually mean it, then the speech act is – in Austin's terminology – 'unhappy', 'void' and 'hollow'. Although an insincere *I do* might seal a marriage simply because the words are intelligible to those present, it is not the kind of speech act Austin considers 'legitimate'. He proceeds to exclude all such examples – where language is not filled with the appropriate 'thoughts and feelings' – from his theory of language.[4] This includes artful and literary language, and a much-quoted passage in *How to Do Things With Words* reads:

> [A] performative utterance will *in a particular way* be hollow or void if said by an actor on the stage, or if introduced in a poem, or spoken in soliloquy. (p. 22)

Austin calls such insincere or merely artful utterances 'etiolations'. They are like plants that grow without light. They are weak and pale in color, deprived of natural vigor and parasitic upon what Austin calls the 'normal use' of language; that is, contexts where speakers and writers mean what they say, where their actions are accompanied by the necessary 'thoughts and feelings'.

Derrida responds to this by arguing that the very fact that signs can be repeated, can be taken out of context and are available for citation in new contexts – whether heartfelt, serious or as a joke – is what language is all about. What matters is that we can perform linguistic actions successfully even when we don't have the requisite 'thoughts and feelings'. Thus, deception – saying things we don't mean – is not outside of language, and the fact that we can do it successfully tells us something about language. In other words, deceitful acts and utterances work, because they repeat what is expected in a particular context (see also Chapter 8). Derrida gives the example of a signature that is meaningful only because it is repeatable, and that, depending on the skill of the forger, will remain meaningful even when counterfeit. Thus, Austin's etiolations – deceptions, language as insincere, artful or staged – are instructive examples of how and why language works. Signs – linguistic and other – are able to mean only because they have meant before, because they are repeatable, because they are parasitic on earlier uses and can be cited. Derrida calls this reanimation of signs 'iterability' or 'citationality', and states axiomatically: 'A sign which took place "only once" would not be a sign' (cited in Sturrock 1993: 124).

The complex interplay between repetition and creativity, sameness and difference, which I discussed above with reference to recontextualization, is also central to Derrida's thinking. According to Derrida the neologism *iterability* should be read as combining two roots, two meanings. On the one hand, there is Latin *iter* ('again') and, on the other hand, there is Sanskrit *itara* ('other'). In other words, iterability, just like intertextuality, is just as much about sameness – fidelity and replication – as it is about change, difference and non-identity. The importance of change was taken up by Judith Butler (1997), who emphasizes that every repetition, every recontextualization, every citation is also a resignification, 'a repetition that is at once a reformulation' (p. 87; see also Kristeva [1969] 1980: 66).

These language-philosophical reflections are important to consider in online environments, where not only do we see rich intertextual relations between texts (typical of signs in general), but copy-and-paste applications might create the illusion that it is possible to produce perfect copies of the original. Surely, the photo on my hard drive and the photo I just uploaded on Facebook are identical? Not actually. The very ideas of intertextuality and resignification mean that a perfect copy is philosophically impossible. The photo on Facebook might indeed look just like the photo on my hard drive, but by inserting the image on my own timeline, in the specific context of my online presence on Facebook, I necessarily change its meaning. And again, like anything else in the social world, resignification too comes in different shapes and forms. Not only might we repost and repeat a text, but we can also engage in its creative and deliberate modification.

REMIXES, MASH-UPS AND SPOOFS

Following this theoretical excursus, let us return to digital practices, which – like all forms of signification – rely on the general iterability of signs, and simultaneously afford new opportunities for recontextualization. Texts that generate extensive online user engagement are often described as viral and are colloquially referred to as internet memes.[5] These are ideas, songs, catchphrases or images that spread across the internet in an epidemic fashion. That is, they are passed from person to person like a virus and spread with great speed across vast distances (countries, social groups, networks). Initially, the text will spread slowly, but once a critical mass has been reached, the text will spread very quickly, before tapering off. Figure 5.4 illustrates this pattern by showing the daily views for the diffusion of 'Gangnam Style' from its release in August 2012 to August 2013. The graph shows how the video reached a peak quickly, then a second peak shortly afterwards, before sharply tapering off. Only if we see such a speedy peak-and-decay pattern in daily views would we call the spread of a text viral. By contrast, a steady growth of viewers (non-viral

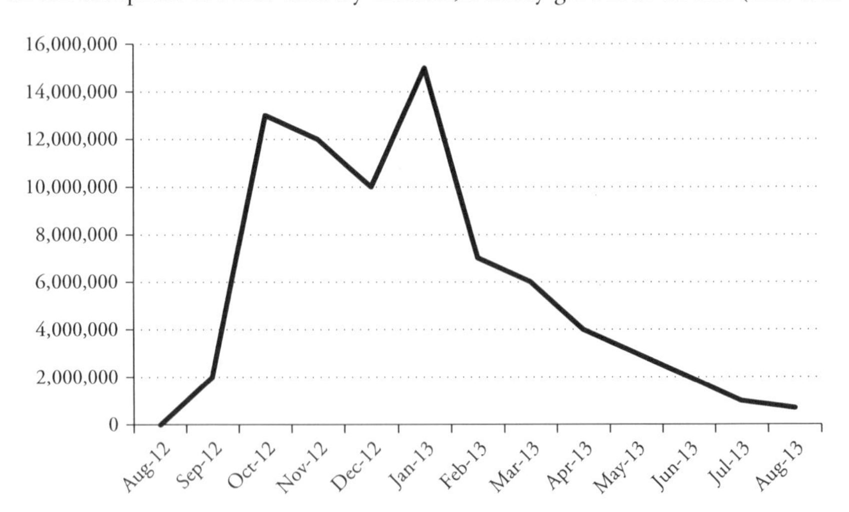

Figure 5.4 Daily views for 'Gangnam Style' (2012–13; based on YouTube video statistics)

diffusion) would create a fairly flat line of daily views, stable and without peaks and declines (Nahon and Hemsley 2013).

Diffusion via personal networks is not new, and word of mouth has always been an important mechanism in spreading information. However, what is new is the speed and scale with which this happens. Instead of days or weeks, a story can now reach thousands of people across the world within hours. In addition to person-to-person diffusion, media coverage and discussions on blogs (including micro-blogs such as Twitter) are important catalysts that help to increase the popularity of online texts across social media platforms. I might, for example, read about a new video on a blog or a news site, recommend the site to a friend and post the video link on Twitter, thus reaching diverse audiences, which, in turn, might further distribute the text (Shifman 2013).

Barack Obama's slogan *Yes we can*, discussed earlier in this chapter, can be used to illustrate processes of entextualization and recontextualization. The slogan featured prominently in the concession speech Obama gave in January 2008, after losing the presidential primary election in New Hampshire to Hillary Clinton. Obama used a wide range of media, including the internet, to get his political message to supporters, and, as noted above, the speech was uploaded on YouTube almost immediately.[6] While many parts of the speech had the potential to become slogans ('there is something happening', 'the power of millions', 'free this nation') and to lead to audience engagement, it was *Yes we can* that was picked up by the live audience (1).

(1) *Obama:* For when we have faced down impossible odds; when we've been told that we're not ready, or that we shouldn't try, or that we can't, generations of Americans have responded with a simple creed that sums up the spirit of a people.

Yes we can.	[Audience cheers]
Yes we can.	[Cheering continues]
Yes we can.	[The audience chants 'yes we can']
Yes we can (very softly).	[The audience continues to chant 'yes we can']

As in the case of the invisible-horse dance in 'Gangnam Style', repetition emerges as a central structural device. Obama's reiteration of *Yes we can* lifts this particular phrase out of the flow of the speech, and the drama of the event – losing the primary but not giving up – contributes to its memorability. Moreover, the slogan is not new but carries echoes of earlier campaigns. Obama had used it already in 2004 in his bid to become a senator, and the Spanish version – *sí se puede* (literally 'yes it can be done') – was the motto of the American farm-worker movement lead by César Chávez in the 1970s. The slogan signals resilience, hope and ethnic inclusiveness – sentiments that defined Obama's campaign.

Obama's speech was picked up by the artist will.i.am (of Black Eyed Peas) in February 2008 and turned into a YouTube music video.[7] While the YouTube video of the speech was certainly popular (with five million views), the music video more than quadrupled the audience (close to twenty-five million views in 2013).[8] The video explicitly acknowledges its source: the text of the song reproduces Obama's speech verbatim, and the black-and-white visuals juxtapose images of the performing artists with those of Obama. Apart from likes and dislikes, the commercially produced music video silences direct audience responses on YouTube: 'Comments are disabled for this video.' The very popularity of some videos can elicit abusive comments, and turning off the comment function is not necessarily unusual (as also noted above for 'Kony2012'). However, disabling the comment function does not stop audience engagement, and the video as well as the slogan itself soon inspired further recontextualizations. This included the music parody 'Yes we can reworked' (2008). Here the musical score remains intact, but the text of the speech is rewritten and distorted images of puppets are used as singers: 'It was indeed written on the back of a beer mat and declared the way to the bar by the station. Yes we can.'[9] Another example is the video 'john.he.is', which uses the same style of black-and-white collage, but instead of Obama shows images of his rival John McCain and announces a 'time of war'.[10] Others, perhaps reminded of their childhood, combined images of Obama with that of the cartoon figure Bob the Builder, whose trademark slogan is 'Can we fix it? Yes we can!' (Figure 5.5). The phrase also appears at the end of an award-winning advert for the legalization of marijuana: 'President Obama, it's time for legalization. Legalization. Yes we can' (2009).[11] This call, in turn, was answered by the mock-documentary 'Yes we can make weed legal' (2012), which showed Obama as a delinquent drug addict bent on destroying the country.[12]

While many recontextualizations of *Yes we can* were – and continue to be – produced in the United States, the phrase also caught on globally. In 2009, the Japanese anime series *Gintama* screened an episode in which the main characters are sick with flu. One of them is Katsura, who has been turned into a Black man by the virus and is now a lookalike of the actor Will Smith. Katsura's name is changed to *Ill* Smith for the episode, a pun on the Japanese second-language pronunciation of the name 'Will'. In one scene Ill Smith/Katsura encourages the others to overcome their sickness Obama-style: 'yes we can' (spoken with a heavy Japanese accent). A short extract from the series was posted on YouTube in 2011 by a user from the Netherlands. The clip not only illustrates the recontextualization of *Yes we can* into a Japanese manga series, but also recontextualizes a fragment of the copyright-protected series in the public sphere of YouTube. The video, subtitled and intended for consumption by non-Japanese audiences, also employs a sociolinguistic practice that can accompany intertextuality, namely metapragmatic reflexivity; that is, a reflexive awareness of what one is doing as one recontextualizes certain linguistic forms and phrases. However, reflexivity is rarely so overt and explicit, declaring the parody to the audience rather than merely performing it (Figure 5.6).[13]

Two years later (2013), when the NSA surveillance scandal dominated news headlines around the world, the phrase was recontextualized once more, this time along with Shepard Farrey's Obama/HOPE portrait, which was created during Obama's presidential campaign in 2008. Following the revelations of Edward

Figure 5.5 Barack the Builder (both 2008; http://www.youtube.com/watch?v=oGjyikboljw; http://www.youtube.com/watch?v=3Btqvc8Gr8k)

Snowden, several mash-ups of the HOPE poster appeared on the internet with the caption 'yes we scan'. The image and caption were also used on 'real-life' placards carried by demonstrators across the world, illustrating again the entanglement of online and offline (Lapidos 2013). In March 2014, another version circulated online (and offline). This poster recontextualized the HOPE design, but replaced

Figure 5.6 Manga video fragment: 'Katsura is Ill Smith' (2011; https://www.youtube.com/watch?v=MuylJDEZN4M)

the image of Obama with that of the Turkish prime minister Recep Tayyip Erdoğan. The slogan was changed to 'yes we ban', denouncing internet censorship in Turkey (Figure 5.7). And so the chain of recontextualizations – online and offline – continues.

Betsy Rymes (2012) calls this the *Yes-we-can-story*, a potentially infinite intertextual series that, as illustrated in Figure 5.3, surrounds the center-piece of the original speech, and each of which, in turn, generates its own audience engagement, a bit like Russian dolls. *Yes we can* is an example of what Debra Spitulnik (1996: 166) has called public words: 'standard phrases such as proverbs, slogans, clichés, and idiomatic expressions that are remembered, repeated, and quoted long after their first utterance . . . [they are] the stuff of popular culture'. Drawing on Benedict Anderson's (1991) discussion of nations as imagined communities, Spitulnik argued that such mass-mediated discourses can establish a sense of togetherness among listeners and viewers, who – like the citizens of a nation – might never actually interact with one another. This can result in the formation of discourse communities as individuals read, see or hear similar texts, and thus come to share resources for meaning-making (see also Chapter 8 on virtual communities).

The global infrastructure of the internet allows audiences to take globally circulating texts and to reframe them at different scales, and while some recontextualizations are intelligible to global audiences, others are deeply local. The global hit 'Umbrella' (released in 2007), by the Barbardian artist Rihanna, attracted a large number of cover versions and parodies on YouTube. Among them, a version dubbed in Bavarian dialect, discussed in detail by Jannis Androutsopoulos (2010),

Figure 5.7 From *yes we can* to *yes we scan* and *yes we ban*

provides a good example of recontextualization-as-localization.[14] The parody, produced by a young Austrian with a self-confessed passion for 'silly songs' (*depperte lieder*), takes its cue from the wordplay *umbrella/an preller*.[15] The latter is a Bavarian expression, which sounds similar to the English title, but means 'to be drunk'. The video retains the original musical score; yet the lyrics have been replaced in their entirety. They are now in Bavarian dialect (delivered by a male voice) and are not about love, but about getting drunk. This is accompanied by a collage of images, many of which draw on local stereotypes of what it means to be a true Bavarian. And as in the case of the *Yes-we-can-story*, the recontextualizations of the originating text prompt their own recontextualizations. Thus, in a tagged response video to 'An Preller', the satirical and humorous lyrics of the video are performed in a serious and melancholic key. Rather than a boisterous drinking song, we are listening to a sad lament about binge-drinking.[16]

Androutsopoulos (2010) discusses such remixes using the metaphorical term 'guerrilla double-voicing'. Double-voicing is a concept introduced by Bakhtin ([1929/1963] 1984: 189ff.) to describe the way intertextuality works. The concept is similar to Derrida's notions of iterability and citationality; that is, it refers to the citation of the voices of others in our own speech or writing. Double-voicing exists in two modes: uni-directional and vari-directional. In uni-directional double-voicing, the speaker uses the 'voice' of someone else in a way that is broadly consistent with the intended meaning of the original (the *Yes-we-can-story* provides many such examples where the phrase is used to express an attitude of hope and change). In vari-directional double-voicing, the speaker 'introduces into that discourse a semantic intention that is directly opposed to the original one' (Bakhtin [1929/1963] 1984: 193). Thus, 'An Preller' recasts a romantic love song as a drinking song, which in turn is recast as a melancholic ballad. It is vari-directional double-voicing that reminds one – metaphorically – of guerilla tactics. There is surprise, spontaneity and unpredictability as well as a lack of regulation, subversion and transgression.

Such, typically ludic, inversions of existing texts are examples of parody, and, like all guerrilla tactics, they are an important tool of political critique. In China, spoof videos, which remix official media products, such as movies or political propaganda, are called *ègǎo* (which can be translated as 'malicious-work', also rendered variably as *e gao* or *e'gao*; Yu 2007; Meng 2011; Link and Xiao Qiang 2013). An example is Hu Ge's parody of Chen Kaige's high-budget movie *The Promise* (2005), a lavish fantasy epic about love, violence and power. The online spoof video takes its cue from the beginning of the movie when Qingcheng, the female lead and an orphan, steals food because she is starving. The spoof – titled 'The Bloody Case that Started from a Steamed Bun' (2006) – develops a story line about a murder case that was caused by a stolen *mantou* ('steamed bun'). The video combines various types of text. Snippets from the original movie are inserted into a mock version of the TV program *China Legal Report*, the narrative is interrupted by bogus commercials, and a mix of various musical genres provides the soundtrack. These range from Chinese love songs to *The Matrix* and revolutionary songs such as the 'Ode to Red Plum Flower'.[17] Hu Ge articulates pertinent social issues in his remix video. Among those are the precarious socio-economic status of rural migrants in urban China, the underground sex industry, and conflicts between city inspectors and street vendors. Upon downloading Hu Ge's video the following text appears: 'The clip you will see is the product of my self-entertainment. The content is purely fabricated. It is for individual entertainment only and dissemination is forbidden' (Gong and Yang 2010: 14). This written disclaimer – a play on familiar copyright warnings – is a parody itself. It subverts authoritative discourse (copyright law), downplays the socio-political critique of the video ('just for fun'), and commands audiences, tongue in cheek, *not* to share and copy the video.

While the *Steamed Bun* parody can be traced back to a particular author, other *ègǎo*s have evolved in truly rhizomatic fashion; that is, like rootstocks without a clear point of origin and sense of a privileged direction, but rather expanding and moving in multiple ways (Deleuze and Guattari 1987). An example of this is the *grass-mud-horse*, a fictitious animal species, which occurs in various guises across the internet. The name in Chinese (cǎo-ní-mǎ, 草泥马) is a pun, and when written with different characters and pronounced with different tones the innocent-sounding name turns into a profanity ('fuck your mother'). The grass-mud-horse *ègǎo* emerged in response to the 2009 government campaign to clean up 'low' and 'vulgar' internet content. It has since turned into a 'virtual carnival', crisscrossing offline and online (Meng 2011: 45; on digital carnivals see also Chapter 8). There exist fake encyclopedia entries, music videos and mock documentaries, comics, and even real-life cuddly toys. The dissident artist Ai Weiwei, famous for his rebellious art, has integrated the grass-mud-horse in various ways into his work. He has published a naked self-portrait in which he covers his genitals with the cuddly grass-mud-horse toy (2009), released a video of himself singing the grass-mud-horse song (2011) and created a cover version of 'Gangnam Style' called 'Grass-Mud-Horse Style' (2012; see also Link and Xiao Qiang 2013).[18]

SPECTACULAR SOUNDS: LANGUAGE AND DESIRE

Parody is an important form of intertextual engagement. However, not every audience response that goes against the intentions of the author is deliberate parody or willful distortion. Derrida's argument for the fundamentally iterative nature of all signs and sign systems makes two important points about meaning. Firstly, as discussed above, only by drawing on signs and texts that have been used before are we able to communicate with others. If we were to create words afresh each time we speak, no one would be able to understand us. Secondly, and only seemingly contradictorily, the iterability of language, which allows us to communicate, simultaneously carries with it the possibility of communicative failure. Why would that be? Simply because the earlier uses of any sign are countless and illimitable, and consequently, the meaning of a sign is never singular. This philosophical point also underpins Eleanor Ochs' notion of indirect indexicality and Jane Hill's notion of mock language, discussed in Chapter 4.

Derrida ([1968] 1982) introduced the neologism *différance* to capture the fact that the meanings of signs are not only different from one another, but also infinitely indeterminate. When spoken *différance* sounds just like the French word for 'difference', *différence*. It is a morphologically perfectly acceptable homophone that is distinctive only in its written form. Derrida coined the noun *différance* to capture the fact that the French verb *différer* means not only 'to differ' but also 'to defer' or 'to postpone'. In other words, meaning is not only located in the difference between signs (e.g. the sound sequence /bet/ is able to carry a meaning different from the sequence /pet/ because of the difference between /p/ and /b/), but is also always deferred, changing and never fully present. Derrida's argument has far-reaching consequences. If all signs – because of their complex histories – are multivocal, without singular meaning, then this means that authors can never control how audiences will understand their texts.

In other words, the meanings I, the author, intend to communicate – Derrida calls this *vouloir-dire*, that is, that which I am 'wanting to say' – might not be what is heard. Meaning is not anchored in the author's voice, but emerges dialogically as audiences respond to what they see and hear. The ways in which we come to understand signs can be shaped by what John Gumperz (1982, 1992) has called contextualization cues. These non-referential signs do not fix or tie down meaning – as this would be philosophically impossible – but *frame* the interaction in particular ways, as serious, as playful, as instruction or as entertainment. However, what counts as a contextualization cue is never given from the outset. It is emergent ('brought about'), negotiated and, just like meaning in general, ultimately, unpredictable and uncontrollable (Blommaert 2005: 42–3). And given the fundamental multimodality of all communication, anything can become such a cue: a word or an accent, a gaze, a pause, a hand movement, the way we carry ourselves, even the way we do our hair.

Among YouTube's vast repository of user-generated content are videos that put language itself on display. This includes songs, poetic recitations and performances, as well as language lessons. The last of these are the focus of the following discussion. When learning a new language, mastering the inventory of sounds can be a challenge. The so-called click consonants, which are present in many languages of southern Africa – among them isiXhosa – are perceived as difficult by many second-language learners, and teachers usually dedicate time to teaching them.[19] As typologically rare sounds, clicks are not just experienced as 'difficult', but also as exotic and even pleasurable, acoustically salient and capable of carrying an expressive load quite unlike other, more familiar consonants. Click consonants have a fairly long history of popular mediatization. Miriam Makeba's performance of 'Qongqothwane' – also known as 'the click song' – introduced these sounds to the global stage in the 1960s, as did the movie *The Gods must be Crazy* twenty years later. More recently Russel Peters, an Indo-Canadian comedian, performed his '!Xobile' skit – about a South African whose name begins 'with a click' – before equally global audiences.[20]

A YouTube search for the keywords *isiXhosa* and *clicks* yielded around six hundred results (April 2014). Many of these were uploaded by visitors to South Africa who would ask someone to perform and explain the clicks in front of a running camera. The top video in terms of view count was an instructional one titled 'Xhosa Lesson 2'. The creator of this three-minute video is XhosaKhaya, a South African blogger, tweeter and commentator (Khaya Dlanga in actual life). The video is shot at a close angle: the face of the teacher – XhosaKhaya – fills the entire frame, and he speaks directly and intimately to the student (Figure 5.8). There is a sense that he is right there, almost jumping out of the frame. The very aesthetic of the video positions the audience not as spectators, but as participants in the lesson.

The video was uploaded in 2009, and in April 2014 it had over 400,000 views and more than 600 comments. These were posted predominantly by viewers from the United States, followed by Europe, the United Kingdom and Australia/New Zealand. Such a global commenting pattern is fairly unusual, as the majority of YouTube videos – unless they have 'gone viral' – attract local rather than global audiences (Brodersen et al. 2012). Responses are mostly in English, which is also the main language used in the lesson. However, there were also isolated responses in German, Spanish, Russian, isiXhosa and Dutch.

Viewers are highly appreciative of the video and the language it displays. They describe isiXhosa as 'awsome', 'beautiful', 'fun', 'too cool' and 'perfect beatbox language'. Negative comments are rare, quickly debunked by others or deleted as offensive. At times, comments are cast as playful textual performances, mimicking the auditory sensation – see examples (2)–(4).

Figure 5.8 'Xhosa Lesson 2': screenshot (reproduced with permission from XhosaKhaya)

(2) click click clickity click

(3) click it or ticket

(4) clik clik clonk clonk cloonk

Metalinguistic discussions about the phonetics of clicks are a common theme. They lead to several mini-instructional moments, where those versed in linguistic terminology get a chance to show off their specialist knowledge – see example (5).

(5) A: I'm trying so hard but still can't get it. Do you inhale or exhale when doing the click?
B in response to A: Clicks are non-pulmonic sounds, that is, they don't involve the lungs. Try hold your breath and form these sounds, then try to slowly add the vowels after that.

Neither the appreciation of clicks, nor the playful performances, nor metalinguistic comments are surprising; after all, click consonants are the topic of the video and viewers are encouraged to practice as they watch. However, not all comments focus on clicks and their articulation. There is an additional set of comments that focus on the teacher and make him the star of the performance. Considering the intimate aesthetics of the video, the camera angle and the close-range visuals, this is less surprising than one might have otherwise thought: the visual aesthetics of the video direct the gaze toward the performer himself and his physical characteristics become the object of attention – see examples (6)–(8).

(6) Your smile and eyes are amazing. Kisses from Poland:)

(7) You seem really happy, I love your face in general:D

(8) I agree. He IS beautiful. Distracted linguist here.

Yet the performer is not just any person; he is a Black man – and this carries its own meanings and fantasies. Many of the comments not only compliment the performer, but also reproduce, and thus recontextualize, racial stereotypes long associated with Africans. Examples of this are comments that focus on the performer's teeth – see examples (9)–(11).

(9) I kept staring at your teeth the whole time. o___o so beautiful

(10) Btw your teeth are really wooow!

(11) 16 people got blinded by his teeth:)

The exaggerated whiteness of African teeth is a classic feature of Blackface, the depiction of Blackness by White actors, and a long-standing stereotype: 'The African Negro has beautiful, pearly teeth, clean, white and perfect' (*The Spokesman-Review*, 10 February 1908). It is an example of what is sometimes euphemistically called 'benevolent racism' (an obvious oxymoron). That is, viewers construct racial difference on the basis of physiology, but then fill the perceived difference with positive, rather than negative, meanings. Thus, Africans are said to possess not only beautiful teeth, but also beautiful smiles and a happy and carefree disposition; they are said to be good athletes, dancers and singers (Wall 2009; Pieterse 1992). By foregrounding, the performer's teeth, as well as his eyes, his hair and his smile, the audience reframes the video, and their comments 'hail' race into being. They are **interpellations** in Louis Althusser's (1971) sense, calling the performer's physical Blackness into being without ever mentioning race directly.[21] The video is no longer just a language lesson, nor is it simply a video of an attractive man filmed at an intimate angle. It is a video of an attractive *Black* man and morphs into a spectacle of racialized exotic beauty.

Comments about the performer's tongue push the reframing even further, and suggest that viewers might not occupy the same interactional frame as the producer of the video. Talking about tongue positions and illustrating such positions is integral to many phonetic classes, and in this educational context XhosaKhaya explains and illustrates tongue positions appropriately: 'see how I place my tongue at the roof of my mouth, it's like, its not totally flat, is just the tip of my tongue . . . not completely flat'. However, the tongue is also essential to many sexual acts, ranging from French kisses to cunnilingus. A large number of comments are overtly sexual, not just appreciating the performer's attractiveness, but also commenting on his assumed sexual prowess and skill – see examples (12)–(15).

(12) I am a married girl and I fell in love with your tongue:*

(13) i bet you eat good vagina

(14) Professor Cunnilingus teaches the alphabet . . .

(15) I bet Xhosa are good at oral sex with women.

Just like certain physical attributes, sexual desire has long been projected onto Africa and Africans, who have been constructed as symbols of lust and boundless sexuality (Fanon [1952] 1986). The South African psychologist Kopano Ratele (2004) describes this mix of desire, taboo and fear, which reconfigures race through the lens of sexuality and desire, as 'kinky politics'.

While a sexualized racial gaze is clearly evident in several of the comments, the audience responses to 'Xhosa Lesson 2' also link to broader intertextual discourses about unknown languages, exoticness and desire (Piller and Takahashi 2006; Takahashi 2013). The short dialogue in example (16) occurred on the comment pages to another video, 'Clicking with Xhosa', in which an isiXhosa tour guide performs a tongue twister (filmed by two American tourists; the video has since been removed). Here the clicks themselves are experienced not only as exotic or interesting, but as sexual and sensual.

(16) A: thats was HOT!lol

B: @A Just wondering . . . how exactly? I mean no disrespect I am glad languages are getting this kind of attention but I am just wondering how it was sexually attractive.

A: *@B* to tell u tha truth.i have no clue.it jus is lol (2010)

Building on Wendy Langford's (1997) psychoanalytic perspective, one can argue that foreign languages allow for the expression of so-called alter relationships; that is, they create interactional spaces where the normal, the routine of the everyday, is suspended. Moreover, since the language is only heard, but not, or only partially, understood, it allows fantasy free rein, and meanings multiply. This is famously depicted in the 1998 movie *A Fish Called Wanda*, where the female lead character becomes sexually aroused merely by being spoken to in an unknown language (Italian and Russian).

In the case of Obama's slogan *Yes we can*, as well as Rihanna's song 'Umbrella', recontextualizations involved not only comments on the YouTube representation of the originating text, but also countless video and image responses, combining speech, print and visuals in creative ways. The video 'Xhosa Lesson 2' did not elicit the same kind of multimodal engagement. No response videos were posted and audience engagement was solely in the form of written comments. However, just as in the case of the more multimodal recontextualizations, audiences resignified the video in various ways. While some engaged with the video as a language lesson, others went well beyond this and foregrounded the performer's bodily presence, ranging from a general appreciation of his attractiveness to racialized and highly sexually charged comments. These audience responses show how difficult it can be to 'control' meaning, and how audiences freely recontextualize the texts that are before them.

CONCLUSION: AUTHOR-AUDIENCES AND THE MOBILITY OF TEXTS

This chapter has looked at digital spectacles and the ways in which they are embedded in various forms of creative audience engagement. Audiences are always both spectator and author. They watch and listen, and they recontextualize the originating text by talking about it, imitating it, repeating it, subverting it and parodying it.

Intertextuality is an important concept for understanding creative agency as something that does not emerge *ex nihilo*, but rather involves the recombination and remixing of existing forms and meanings. Intertextuality further draws our attention to the fundamental iterability of language, a central theoretical point in Derrida's oeuvre. From this perspective, meaningful communication requires the repetition of existing signs, but simultaneously allows us to go beyond mere imitation by acknowledging the essential multivocality, and indeed ambiguity and openness, of all signs. In the case of 'Xhosa Lesson 2', for example, an instructional video about how to pronounce clicks turned into a spectacle of exotic beauty and desire. This shows that authors can never exert full mastery over meanings and that audiences will always, in some way, 'do their own thing'. Yet, at the same time, neither is the audience the sole and unconstrained producer of meaning. Rather, the understanding of a text is guided by contextualization cues – such as the close camera angle and the intimate visuals – as well as socio-historical intertextualities, in this case a long history of racially defined notions of desire and attractiveness.

The notion of iterability is important for the theme of mobility, which was discussed in Chapter 2. Not only are new media increasingly mobile in terms of their materiality, they also facilitate the rapid mobility of texts across different scales, from local to global and anywhere in between. And as texts move from one context to another, they change and transform. If we want to understand this process we need to follow their movements and uncover the traces of past uses (using, for example, the follow-the-text approach discussed in Chapter 2). Understanding intertextuality thus requires a historical mindset, a return to philology. It also requires an ethnographic mindset, that is, close attention to the details of production, context and audience reception in order to write nuanced 'ethnographies of [mobile] texts' (Blommaert 2008: 12ff.).

The theoretical focus in this chapter has been on what is generally referred to as poststructuralism, with Derrida as a central figure. In the next chapter I will turn more strongly to Bakhtin, who has been described as a poststructuralist *avant la lettre* (Young 1985/6). However, I will also continue the discussion of Derrida since his view on speech and writing is important for linguists to consider.

NOTES

1. I follow Hodges (2011: 11) in using *originating* rather than *original*: 'any text is always comprised of prior texts and it is misleading to speak of an "original" context' and text.
2. A concept similar to recontextualization is Iedema's (2003) notion of 'resemiotization'.

3. Seeing repetition (and thus memory) rather than rule-driven generative operations as fundamental to language is evident in the work of Hymes, Becker, Bolinger and Hopper (discussed in Tannen 2007: 48ff), and more recently Taylor (2012).
4. Austin distinguishes between 'misfires' and 'abuses'. If I say *I do* in front of someone who has no authority to seal a marriage, or if I say it to my cat or goldfish, then the speech act 'misfires', as the external conditions necessary to seal a marriage are not given (the marriage officer needs to be authorized; we cannot usually marry our pets). However, if I say it to my human groom or bride in front of a priest, but don't mean it and am already plotting the divorce, then this is 'abuse', as I don't have the required 'thoughts and feelings' to make this a 'happy' performative.
5. The term *meme* goes back to Richard Dawkins' book *The Selfish Gene* (1976), and was intended as the cultural equivalent of *gene*. A meme is thus an idea or text that gets replicated and mutates in the process. The site knowyourmeme.com documents a wide range of internet memes, including videos, images, catch phrases and web celebrities. Many internet memes have been recontextualized in offline popular culture and have become cultural reference points.
6. http://www.youtube.com/watch?v=Fe751kMBwms.
7. https://www.youtube.com/watch?v=jjXyqcx-mYY.
8. Both videos spread virally, showing the typical peak-and-decay pattern. While the music video reached immediate popularity in 2008, and peaked again at the time of Obama's inauguration (January 2009), the video of the speech showed only a small peak in 2008, and spread mainly in January–February 2009.
9. https://www.youtube.com/watch?v=qmMEFYzxmgs.
10. http://www.youtube.com/watch?v=3gwqEneBKUs.
11. http://www.youtube.com/watch?v=C0mEDE_w1xo.
12. http://www.youtube.com/watch?v=QAy9a6f41LM.
13. The episode (165) dates back to 2009 (http://myanimelist.net/forum/?topicid=97688).
14. http://www.youtube.com/watch?v=icmraBAN4ZE. A translation of the lyrics is given in Androutsopoulos (2010).
15. Information about the creator of 'An Preller' can be found on his homepage (http://www.schwappe-productions.de.vu).
16. http://www.youtube.com/watch?v=P6rzbBv7p1o.
17. A version of the spoof video – without subtitles – is available at: http://www.youtube.com/watch?v=AQZAcT1xaKk. The original remix was not distributed via YouTube, but on blogs and other websites (such as sina.com).
18. http://www.youtube.com/watch?v=Nsk-KgD0aOM; www.youtube.com/watch?v=4LAefTzSwWY&feature=youtu.be.
19. IsiXhosa has three basic clicks which are represented by the symbols <x> (lateral), <c> (dental) and <q> (palatal).
20. http://www.youtube.com/watch?v=Yj-1kp777NM. For maximum visual effect the name *!Xobile* combines two separate orthographies. In isiXhosa lateral clicks are indicated by the letter <x>; the orthographic conventions of Khoisan languages use an exclamation mark to represent a post-alveolar click. If one were to pronounced *!Xobile*, one would have to articulate a post-alveolar click followed directly by a lateral click. This, however, is structurally impossible, as in actual languages clicks are always separated by vowels or nasals (personal communication, Matthias Brenzinger). Both the orthographic bricolage and the sound structure of the name are thus entirely imaginary.
21. There is, of course, one important difference from Althusser's example of a policeman calling 'hey, you there!', and the person thus addressed turning around. In 'Xhosa Lesson 2', the subject does not respond to or acknowledge the interpellation.

Chapter 6

Bakhtin goes mobile

> oh my. someone is tweeting as me. as me. but . . . like, BY me. AS ME, BY ME. WHAT. WHOSE VOICE. BAKHTIN. is that you? no. a-ho, boy. sailor J.
>
> Twitter 2013

INTRODUCTION: ARTFUL INDEXICALITIES

In the previous chapter I have argued that intertextual relations are fundamental to the way meaning is created. This chapter continues this line of thinking by drawing more strongly on the work of the Russian language philosopher Mikhail Bakhtin.[1] While work on intertextuality has typically focused on what Zellig Harris (1952) has called 'language beyond the sentence', this chapter uses Bakhtin's notion of heteroglossia to develop a perspective that also allows us to work below the sentence. In other words, intertextual relations also hold for the building blocks of language, that is, individual morphemes, sounds and their graphic representations.

Working at a level below the sentence is familiar territory for sociolinguists, who have long studied – through the lens of variables and their variants – different ways of expressing the same grammatical function or the same phoneme. Sociolinguistic work has shown that linguistic variants index, that is, they point to social and cultural meaning (Silverstein 2003; see also Chapter 4). Consider, for example, the articulation of word-final velar nasal [ŋ] in words such as *going* or *reading*. Clipped, non-standard pronunciations that change the velar to [n] are not neutral, but can index a particular type of speaker: casual and easy-going, but also lazy and uneducated (Eckert 2008). These meanings are cultural construals based on the contexts in which these pronunciations are usually heard; that is, they connect the utterance in the here and now to prior uses, and thus establish intertextual chains of meaning. However, Derrida's notion of *différance* – discussed in the previous chapter – reminds us that the meanings of signs are never closed. They are, on the contrary, always in flux, open and contested. The inherent multivocality of all signs defies the cataloguing of their indexicalities and new meanings can always surface.

When discussing digital language, our focus is usually on writing, which remains, multimodal affordances notwithstanding, a core mode of online interaction. The affordances of writing will be discussed in the first section of this chapter, which takes a look at the old speech–writing binary in linguistics. Much of what is available online – news reports, government sites, health information, educational material – follows the standard norms of orthography and grammar, and language use is fairly regulated. Other types of online engagement – most notably interactive chatting and texting, as well as social networking – are weakly regulated (Lillis 2013: 124ff). They allow for experimentation and the creative use of semiotic resources. In this chapter and the next, I will look at written language use in such weakly regulated spaces. In *Always On*, Naomi Baron (2008: 6–7, 169) argues that because we write more than ever before, an inattention to *how* we write has crept in, a sense of 'linguistic whateverism'. People, she argues, no longer care whether they spell correctly; they write on the fly, with little attention to grammar, word choice and punctuation. Is this actually true? I will argue that, although standard norms are disregarded, informal digital texts are often highly crafted and indeed artful, displaying complex linguistic skills rather than lack of attention. This chapter introduces three concepts that are central to understanding language as artful: heteroglossia, performance (as well as performativity) and stylization.

SPEECH AND WRITING

Thinking in binaries is a common feature of everyday as well as academic discourse: *man* vs. *woman, natural* vs. *artificial, clean* vs. *dirty, haves* vs. *have-nots* (Chapter 3) and, of relevance here, *speech* vs. *writing*. Binaries can be analytically useful as they focus our attention on stark differences between opposites. However, they also obscure much of the ambivalence and indeterminacy that characterize social life. Moreover, they are often 'violent' in that they establish hierarchies and take one term (*man, natural, clean, haves, speech*) as superior to the other (Derrida 1974).

Generally, Western linguists (e.g. Ferdinand de Saussure [1916] 2013) have seen the spoken word as the true locus of language. This phonocentrism – a term used by Derrida – goes back to philosophers such as Plato, Socrates and Aristotle who critiqued writing as deceitful, promoting forgetfulness and being incapable of true dialogue and interaction (Harris 2000; see also Chapter 2).[2] When linguistics emerged as a discipline in the early twentieth century, the primacy of speech, and the secondary nature of writing, soon became dogma: it not only encouraged linguists to study societies that did not have written traditions – an important and positive step – but also banished the study of writing from linguistics proper. From the 1960s onwards, sociolinguistics pushed the primacy of speech even further: it is not any speech we should study, but informal, everyday speech (the vernacular).[3]

However, writing did not entirely disappear from the broader research agenda in the social sciences. In the 1980s and early 1990s, Eric Havelock (1982), Jack Goody (1987), Walter Ong (1982) and David Olson (1994) argued that the introduction of writing into previously oral societies has far-reaching consequences. Writing, according to their argument, enables more complex and abstract thought (by freeing the memory), allows greater precision when transmitting information, and increases metalinguistic reflexivity. This view is similar to the perspective of determinism as discussed in Chapter 3: literacy is seen as a technology that triggers, or even 'causes', a fundamental social and cognitive change. Although these – highly controversial – views were largely discredited in subsequent ethnographic work, which showed that social context shapes literacy rather than the other way around, they nevertheless created a new interest in writing as a symbolic system in its own right, *different from, but not subordinate to, speech*. And in order to understand whether there might exist systematic linguistic differences between speaking and writing, linguists – especially those in stylistics or literary linguistics – compared spoken and written texts.[4] Although some regularities were found, results became muddled as soon as researchers added genre to the mix; that is, texts that frame contexts in particular ways and are produced for specific purposes and audiences. Thus, a spoken lecture would differ more from an informal spoken conversation than from a written, academic text. It soon became clear that it is impossible to identify a single set of features by which *all speech* could be distinguished from *all writing* (Akinnaso 1982; Chafe and Tannen 1987; Biber 1991; Barton 2007). A way out was to approach speech and writing from the perspective of prototypicality: an informal conversation between friends is seen as an ideal type of speech; an academic article an ideal type of writing. These ideal types are then positioned as poles on a (multidimentional) continuum (Figure 6.1).

While the idea of a continuum – which allows for gray zones and fuzzy cases – is preferable to that of a strict binary, it nevertheless recasts the old binary (and the hierarchy it implies). A written text that shows characteristics such as 'involvement' and 'informality' would be described as *speech-like*, and as such would warrant study by, especially, sociolinguists who continued to foreground 'the vernacular', that is, maximally unmonitored spoken language, as their main area of interest. The reading out of a written newscast, on the other hand, although oral in its modality, would be too *written-like* to be of real sociolinguistic interest. Similarly, work on digital writing has concentrated largely on texts that were considered to be speech-like, especially chatting and texting. Linguists have described such texts as 'typed conversations' or 'fingered speech', and have argued that they 'have much in common with face-to-face conversations' (Vandekerckhove and Nobels 2010: 658; also Dorleijn and Nortier 2009). Certain forms of writing are cast as akin to talk and thus worthy of sociolinguistic study. This metalinguistic framing is echoed by digital writers themselves: people report that they 'chat' and 'gossip' online, that they 'shout', 'yell'

Speech(-like) ------------------------------------	**Writing(-like)**
Informal	Formal
Unplanned	Planned
Dialogic	Monologic
Private	Public
Fragmentation	Integration
Involvement	Detachment,
Present	Absent, etc.

Figure 6.1 The speech–writing continuum

or 'scream' (Jones and Schieffelin 2009). Others have positioned digital language in-between the two modalities, that is, as a hybrid form, which combines stylistic features of speaking and writing (Crystal 2006). This interpretation too recasts the original binary, because digital language can only be in-between if the poles of speech and writing exist in the first place.

The work of Derrida – which was introduced in the previous chapter – provides a rather different perspective. In 'Signature Event Context', Derrida ([1972] 1988) argues that, from a language-philosophical perspective, speech and writing – even in their most prototypical state – share fundamental characteristics. According to Derrida, all language is premised on representation and mediation, and thus on absence. In the case of writing, this is easily illustrated. As I write this text, you the reader are not here, and although you are absent, you are able to read and understand what I write. Yet absence is not merely a question of materiality – of bodies in different places at different times – but is intrinsic to the workings of language itself, and similar absences are apparent when we speak.

Derrida argues that even when we find ourselves in the most intimate physical, bodily presence of others, communication is never the unproblematic communion of two minds. As discussed in the previous chapter, when we communicate we necessarily use linguistic signs that are external to us. They are able to mean only because they have meant before, and their meaning can never be fully determined by our intentions. In this sense, speaking is no more immediate or involved than writing. In other words, all signs stand for something that is absent, and what they represent is never given, determined or *present*, but is the – necessarily deferred – outcome of interactional engagement and interpretation. This is again Derrida's principle of *différance*. Derrida's argument is important for linguists to consider. It implies that there can be no unmediated language, that body-to-body communication does not allow us to learn anything about the nature of language that we cannot learn equally well from the study of written texts (see also the discussion of the mediated interaction order in Chapter 2). Language is language in whatever modality it presents itself.[5]

Derrida's argument is located at the level of language philosophy. Brian Street's book *Literacy in Theory and Practice* (1984) provides a more hands-on empirical program that, like Derrida, puts spoken and written language on an equal footing, both worthy of study and academic interest, both playing a role in the social organization of life. Although it is no longer 'new', this approach continues to be known as New Literacy Studies (henceforth NLS).

NLS scholars critique the way earlier work has constructed literacy (and orality) in monolithic ways and ignored the socio-cultural environment in which diverse forms of literacy are enacted. Consequently, NLS shifts the emphasis toward ethnographic studies of literacy *practices* within specific *contexts* (examples include Scribner and Cole 1981; Heath 1983; Street 1993; Sebba 2007; Baynham and Prinsloo 2009). NLS scholars emphasize the close integration of spoken and written language in many social interactions. Communicative situations where literacy plays 'an integral role' (Heath 1983: 71, 386) are called literacy events. Two friends chatting via instant messenger or someone reading a book on a train would be examples of literacy events involving only written language. At times, literacy events show a close interweaving of spoken and written language. This can be illustrated with an example from my field-notes: two women in their mid-twenties are enjoying a beer at a local food market in Cape Town, while reading and commenting on the text messages one of them receives from a potential male suitor she met the previous night. Each message is first read out, then discussed, and a reply – accompanied by much laughter – is composed. This is followed by the anticipation of the next response. Sociability in this instance involves speech and writing, physical presence and absence, two distinct but overlapping interaction orders.

To argue that speaking and writing are both instantiations of language and that neither is prior to the other, or derivative of the other, does not mean that speaking and writing are the same and that we should disregard their differences. Most importantly, the modalities through which speech and writing are transmitted, namely sound and inscription, bring along medium-specific affordances. Writing is visual, not auditory (a point that will be explored in more detail in the next chapter). It has a noticeable materiality and its production typically involves artifacts (tools): pen and paper, keyboards and screens, fingers and sand, spray can and wall. The modality affects our ability to archive and to create durable texts. I can save a letter, or even a particularly nice text message, but I cannot easily 'save' a compliment someone gave me verbally in the same way (unless I record it; see Chapter 2). And while speaking takes place in a context of *temporal unity* and information-processing happens in real time (Auer 2000), writing allows for *temporal disunity*: it is impossible for you, the reader, to know how long it took me, the writer, to produce this paragraph in the form in which it is now printed. Temporal disunity allows not only for careful, considered composition but also for correction and editing: the eraser and the delete button are just as essential to the writing process as the pencil and the alpha-numeric keyboard. This opportunity for editing encourages reflexivity, which in turn is an important condition for agency and creativity, and shapes digital practices (Joas 1996; see also the discussion on interactional affordances in Chapter 3 and Goffman's 1979 observations on media language, discussed near the end of Chapter 2).

VARIATION AND HETEROGLOSSIA

Language is variable – this phrase articulates a rare consensus view in linguistics. However, there is less agreement on how to interpret and analyze the variation we see. The dominant sociolinguistic approach to the study of variation has been William Labov's variationist paradigm. Central to variationism is the axiom of structured or orderly heterogeneity; that is, linguistic variation is not random or arbitrary, but ordered and patterned (Weinreich et al. 1968; Labov 1972, 2001, 2010). The methodology is mainly quantitative and focuses on the frequencies with which forms occur in particular contexts or are used by certain types of speakers (usually defined according to social demographics). The parameters that govern variation can be *linguistic* (e.g. in English, word-final consonants are often deleted if the following word also begins with a consonant, e.g. *las' month*), *situational* (e.g. consonant deletion is more common in casual ways of speaking) or *social* (working-class speakers might show higher frequencies of deletion than middle-class speakers).

Variationist methods have been used with mixed results in the study of digital language. Some studies illustrate – in line with the axiom of structured heterogeneity – the systematic nature of variation in digital writing. For example, Tyler Schnoebelen (2012) looked at writers' use of emoticons: are they writing them 'with nose' or 'without nose', that is, as:-) or:)? His large-scale study – involving close to four million tweets and more than a hundred thousand authors – identifies regular co-occurrence patterns: those who omit 'the nose' tend to be younger and use more taboo words, more expressive lengthening (*soooooo tired*) and more non-standard spellings.

Work on gender has also made use of variationist methods, and patterns of gendered language use have been identified in online, written language. Thus, studies found that women make greater use of hedges and politeness markers, and adopt a more supportive conversational style (Kapidzic and Herring 2011; Herring 2003). With regard to bilingual speech, John Paolillo (2011) studied code-switching and code-mixing in two different digital contexts: chat rooms, which allow for synchronous interaction; and Usenet discussion groups, which favor asynchronous communication. His results show that bilingual speech was more common in synchronous environments, where interaction takes place in real time and writers experience a sense of immediacy. Such studies illustrate the usefulness of the variationist paradigm in identifying broad, quantitative patterns and remind us that when studying written, digital language there are medium-specific variables to consider (emoticons and other typographic symbols, as well as the technological affordances of different platforms).

Other studies, however, are at odds with the axiom of structured heterogeneity, and encourage us to explore new ways of understanding variation. Sali Tagliamonte and Derek Denis (2008) compared variation in informal spoken Canadian English with variation in instant messaging (henceforth IM) for the same group of participants. Their results for quotative *be like*, one of the variables under investigation, are summarized in Figure 6.2. In the spoken data colloquial *be like* (e.g. 'and then she was like "what are you doing?"') is an emergent norm. The IM data, by comparison, shows a mixture of forms: writers use the colloquial variant, but also make regular use of standard *said* (also *asked, thought* etc.) as well as zero verbs (e.g. 'and he: yes, great!'). The last of these variants is facilitated by the affordances of writing, that is, the availability of punctuation to mark direct speech.[6]

Thus, while the spoken data shows clear patterns that are in line with the principle of structured heterogeneity, the written IM data is characterized by stylistic diversity as well as unexpected co-occurrence patterns. In (1), for example, the writer uses the formal variant *shall* for future reference, combines it with the informal intensifier *serious*, the slang term *jam*, and the medium-specific sound symbolism *aaaaaaaaagh* (communicating excitement by typographically mimicking a scream). The resulting

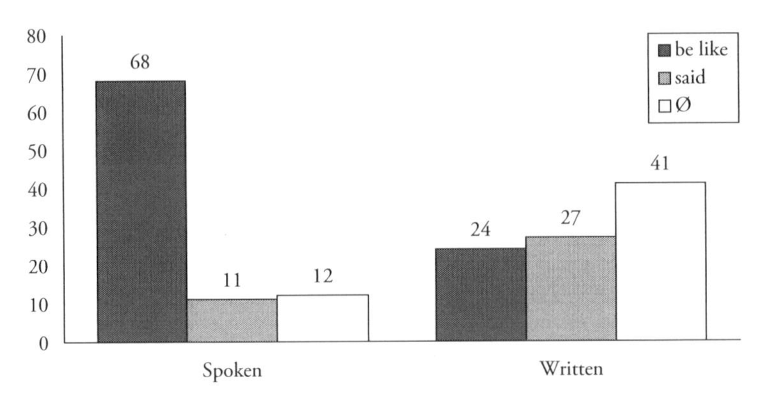

Figure 6.2 Distribution of the top three forms in the quotative system by medium: percentages (Tagliamonte and Denis 2008: 20)

post is stylistically complex and difficult to interpret from the perspective of ordered heterogeneity.

(1) aaaaaaaaagh the show tonight shall rock some serious jam (male, 19 years; p. 26)

Reinhild Vandekerckhove and Judith Nobels' (2010) quantitative study of Flemish chat logs shows similar results to those of the Canadian study. They speak of 'arbitrariness of alternation' and an 'absence of any pattern' (pp. 670f.) – turns of phrase that are at odds with the very idea of structured and patterned language use. Example (2) comes from their data (pp. 670–1): only in turn no. 4 does the writer conform consistently to a particular linguistic system (standard Dutch); all other turns are 'mixed' and include, in addition to Dutch, forms that are associated with colloquial styles of speaking (bold) as well as regional dialect forms (West Flemish, italics).

(2) In this extract the writer is encouraging a friend to send her handwritten letter.

1	**ma** schat toch	**but** darling now
2	doe **ma** *jon* best *wi*	do **just** *your* best *hey*
3	en **kvind nie** erg **da** er *vlekn* zijn	and **I don't** mind **that** there are (ink) *stains*
4	en je mag fouten schrijven	and you are allowed to write mistakes
5	dus denk eens na in *jon bedde* **wa da** je zal schrijven en schrijf **da** 1 keer *up*	so think now in *your bed* **what** you will write and write **that** *down* once
6	en **da** is wel **goe** *wi*	and **that** will be **okay** *hey*

An alternative to Labovian variationism can be found in the substantial and diverse body of work that looks at language not as a rule-governed system, but as a repertoire of signs. In other words, linguistic knowledge is recast as a set of communicative resources that are reworked creatively, rather than predictably, in interaction (Gumperz 1964; also Coupland 2007; Jørgensen 2008; Blommaert 2010; Otsuji and Pennycook 2010; Blackledge and Creese 2010; Rampton 2011). Unlike 'languages' or 'varieties', which are social and ideological constructs ('German', 'Estuary English', 'IM language'), linguistic repertoires are deeply personal: they are 'tied to an individual's life' and reflect a person's experiential trajectory in time and space (Blommaert 2008). With increased physical and virtual mobility, these trajectories diversify' 'bits of language' travel well beyond their originating context and can turn up in quite unexpected places (Blommaert 2010: 28ff.; Pennycook 2012: 17ff.). Repertoires can include different languages, but also different styles of using one language (e.g. regional forms, youth variants, ethnic variants and so forth), and as we speak/write we draw on these forms and combine them in novel ways.

However, this raises an important sociolinguistic question: what guides the selection of forms in particular contexts? Do speakers/writers combine the forms of their repertoire into utterances just as they would assemble items on their lunch tray in a cafeteria? Is the process random or structured? The expression 'cafeteria principle' was originally used by Derek Bickerton (1981) to ridicule those who explained the structures of contact languages as the result of ad hoc language mixing, rather than the systematic and predictable workings of a rule-governed mental grammar. More recently, however, Salikoko Mufwene (2008) and Jeff Siegel (2008) have defended the 'cafeteria principle':

> Although the 'cafeteria principle' is supposed to imply a certain randomness in the selection of features, even the choice of items for lunch at a cafeteria has some constraints: what's available at the time, how much room there is on the tray, what items one recognizes, what goes with what, etc. (Siegel 2008: 79)

The constraints mentioned by Siegel are of a weak nature: they are not systematic rules that apply across individuals, but are rooted, as noted above, in the biography of individuals and in their beliefs about, and preferences in, 'what goes with what'. The workings of the cafeteria principle are always culturally and historically contingent (see also Humboldt on freedom and constraint in Chapter 2).

Consider this example from South Africa: throughout the country, words such as *that* or *this* are pronounced with a full voiced dental fricative [ð], or an approximate fricative. Stop pronunciations occur, but are rare (Rajend Mesthrie, personal communication). Yet in South African digital writing, stop spellings – such as *dat* for *that* – are extremely common (Deumert and Lexander 2013). How did they get there? Probably via global popular culture, that is, written representations of African American English – which uses stops rather than fricatives – in song lyrics and advertising. Similar processes are happening elsewhere, and Mark Sebba (2007: 53), for example, noted with regard to the United Kingdom (where stops are equally uncommon in the spoken language):

> The respelling of <the> as <da>, <d> or <de> seems to be widespread in hip-hop culture and may be becoming common practice in SMS messaging among adolescents.

Linguistic *forms* – written and spoken – are thus mobile. They might *look* (or *sound*) the same in different locations, yet their indexicalites differ from locale to locale. The spelling <da> might signal ethnic and local belonging in Harlem, and a lack of education and refinement at Harvard; among many adolescents and young adults in South Africa (and the UK) it indexes familiarity with global popular culture (see Chapter 7 for further discussion). The use of stop spellings shows that linguistic forms – in this case the graphic representation of a pronunciation feature – can be recontextualized just like texts; they too carry traces of earlier meanings and resignify in new contexts.

While the cafeteria principle is a useful metaphor to think with, Jannis Androutsopoulos (2011) has suggested that Bakhtin's language-philosophical reflections provide a broader theoretical framework for conceptualizing variation online (and offline).[7] A Bakhtinian perspective allows us to understand not only the emergence of regular patterns and conventions, but also mixtures of forms that resist an analysis of orderly – that is, broadly predictable – heterogeneity.

Bakhtin ([1934/1935]1981) argues that language use is shaped by two fundamental forces or processes: centripetal and centrifugal forces. These forces always co-exist, but at different times one or the other might be stronger. Centripetal (toward-the-center) forces bring about linguistic practices that show well-developed norms, and individuals tend to orient their usage toward such norms. It is in such contexts that sociolinguistic description is fairly straightforward and co-occurrence relations can be identified. At other times, however, individuals fill their linguistic 'cafeteria tray' in quite different ways and language use might appear chaotic and patchwork-like. Rather than a particular code with well-established norms being spoken, linguistic forms are reassembled in various, and often surprising, ways. Bakhtin refers to the latter scenario as centrifugal (away-from-the-center), and describes it using the term heteroglossia, literally 'multi-speech-ness' (from Russian разноречие, *raznorečie*).[8] Heteroglossia – a central concept in Bakhtin's thinking – describes the multiplicity of languages, dialects, styles and forms that exist in every society and that provide us with resources for speaking and writing.

Bakhtin approaches heteroglossia from the perspective of intertextuality – or in his terminology, dialogism. That is, as we speak or write in the here and now, we are in dialogue not only with present audiences, but also with those who have spoken before us. That is, we draw on their voices. We speak through their voices and use their words – and the indexicialities they carry – as we create new meanings in new contexts. And as we draw on these heteroglossic resources (existing *social voices* and conventions), we not only enact pre-existing social personae but also develop a sense of self, an *individual voice*, in dialogue with existing voices. This move from citation to individuality is made possible by the principle of double-voicing, which was discussed in the previous chapter. In other words, as we use existing linguistic forms, we insert 'a new semantic intention into a discourse which already has, and which retains, an intention of its own' ([1929/1963] 1984: 189). Or in the words of Sapir, cited in Chapter 2, we give them 'just enough twist' to make them 'our own', and reframe other voices in unique ways in our own speech and writing. Bakhtin summarizes the interaction between social and individual voice in 'The Problem of Speech Genres' as follows:

> This is why the unique speech experience of each individual is shaped and developed in continuous and constant interaction with others' individual utterances. This experience can be characterized to some degree as the process of *assimilation* – more or less creative – of others' words (and not the words of a language). Our speech, that is, all our utterances (including our creative works), is filled with others' words, varying degrees of otherness or varying degrees of "our-own-ness", varying degrees of awareness and detachment. These words of others carry with them their own expression, their own evaluative tone, which we assimilate, rework, and re-accentuate. (Bakhtin [1952/1953] 1986: 89)

The idea of voice is central to Bakhtin's thinking and links to his discussion of polyphony, the possibility of a 'plurality of *independent* and *unmerged* voices' within a text (Bakhtin [1929/1963] 1984: 6; my emphasis). Polyphony does not represent a unified text in which different styles and voices merge seamlessly into something new. Rather, each polyphonic utterance is unique in its combination of linguistic forms and the dialogical relationships, and indeed struggles between different voices, it creates.

Adopting a Bakhtinian perspective, the diversity of linguistic forms we see in examples (2) and (3) appears not simply 'ill-matched' or 'arbitrarily mixed'. Rather they articulate different voices within contextually specific utterances. This creates artful *tensions* and semantic *conflicts*, as well as *harmonies*, in the way they co-articulate with one another (Brandist and Lähteenmäki 2011: 73). The adjective 'artful' is important here. Examples (2) and (3) cannot be understood as casual, maximally unmonitored language, 'the vernacular' in the Labovian sense. They are crafted utterances in which writers draw on their entire repertoire and combine forms for maximum effect. It is important to keep in mind that Bakhtin wrote not only as a language philosopher, but also as a literary scholar, and he described his interest quite explicitly as 'the study of verbal art' ([1934/1935] 1981: 259). Extending Bakhtin's observations, which were inspired by formal literary texts, to everyday spoken and written language certainly requires a shift in the way we view language: not mainly as a social fact characterized by conventions and norms, but also as a source of creativity and art, which produces utterances that are unpredictable and multi-voiced. The remainder of this chapter discusses two concepts that have been central to a sociolinguistic understanding of language as artful and crafted: performance and stylization.

PERFORMANCE AND PERFORMATIVITY

Performance occurs with three different and partially overlapping meanings in sociolinguistics and linguistic anthropology. Performance-1 refers, very broadly, to actual language use, the ordinary things people say and do with language in everyday life. Performance-2 is indebted to Erving Goffman's (1969) work on the dramaturgy of the everyday and describes, more narrowly, how we present ourselves, quite strategically and self-consciously, to others in everyday interactions; that is, the way in which we *routinely* perform social types and personae, and articulate more or less well-defined social voices as we try to manage the impressions others have of us. Performance-3 describes a specially marked and *non-routine* mode of speaking/writing, a mode that evokes the theatrical, displays language as carefully crafted and artful, and invites an aesthetic response – and indeed scrutiny – from the audience. Richard Bauman, whose work has been foundational in this area, defines performance-3 as a way of speaking that shows 'special attention and heightened awareness to the act of expression' (1977: 11). The speaker or writer signals to the audience: 'hey, look at me! I am on! watch how skillfully and effectively I express myself' (2004: 9). The focus on artful speech, reflexivity and aesthetic evaluation is a shift away from past work in sociolinguistics, which has generally focused on performance-1, that is, the study of – presumably – 'natural', spoken, unscripted and unrehearsed language in body-to-body interactions. Bauman's criterion of 'heightened attention', for example, stands in direct opposition to Labov's (1972: 208) 'minimal attention' as a defining characteristic of casual speech and 'the vernacular'. In this chapter I will refer to performance-1 as *language-in-use*, whereas *performance* will be used to refer to both performance-2 and performance-3 Although analytically distinct, there is overlap between the latter two concepts and it can be difficult to keep them separate when analyzing actual data. Both refer to crafted and artful semiotic displays, either as everyday routines (performance-2) or in non-routine moments, which are set apart from ordinary interaction (performance-3).

Examples of linguistic performances in the sense of Bauman (performance- 3) are mass-mediated political spectacles such as Obama's *yes we can* speech (discussed in Chapter 5). The speech itself is an example of what is called 'high performance' (Coupland 2007) or 'full performance' (Bauman 2004). It is a staged and pre-announced event that is planned ahead, rehearsed, temporally/spatially bound and marked off from routine life. However, not everything that has performance qualities is high performance, and performances can also emerge spontaneously in interaction (Hymes [1974] 2004). Everyday performances are neither scheduled nor rehearsed, are often short-lived and transitory, but are nevertheless 'crafted, self-conscious and reflexive' and characterized by 'special attention to and heightened awareness of the act of expression' (Bauman 2011: 708, 710).

In their discussion of how to study an increasingly mobile world, Monica Büscher and her colleagues (2011) encourage us to focus our attention on fleeting and transitional moments in social life, on that which 'slips and slides', comes and goes, 'that which is here today and gone tomorrow, only to reappear the day after tomorrow' (Law and Urry 2004: 403). Moments deserve more careful theoretical attention in sociolinguistics too. They interrupt the routine flow of life, emphasize the agency of speakers/writers vis-à-vis the routines and patterns of everyday life, and highlight the creative possibilities of language – even if just 'for a moment' (see also Li Wei 2011). Short-lived, everyday performances are, for example, a core interest of those working in the small stories paradigm (Bamberg and Georgakopoulou 2008). While big stories are akin to high performances – with a teller-performer, an audience and a structured, even rehearsed narrative – small story research has foregrounded the role brief and fleeting story-moments play in everyday interaction. The performances discussed in this chapter are similar to small stories, and, by analogy, could be termed small performances.

Momentary and fleeting, non-routine performances abound online. Examples (3) and (4) were both posted on the Facebook wall for the group 'How many xhosa's are there on Facebook'. In example (4), the writer plays around with the morpheme */xhosa/* which is commonly used as a noun (referring either to the language, *isiXhosa*, its speakers, the *amaXhosa*, or the place from which they come, *emaXhoseni*, the place of the amaXhosa). It is not usually used as a verb and *ukuthetha isiXhosa*, 'to speak isiXhosa' would be the expected form here. By going beyond the expected, the writer not only creates a spirited tongue twister that celebrates the lateral click as a symbol of Xhosaness, but also shows himself to be a skillful writer who knows the structure of his language so intimately that he can playfully transcend its conventions.

(3) ndingumXhos'oXhos'isXhos'es'ngenoXhosw'ayinjubaq'engengoXhosa:-) (Facebook 2010)
'I am a Xhosa who xhosas isiXhosa which can't be xhosaed by a rascal who is not Xhosa'

In (4), the writer performs a ceremonial isiXhosa salutation, reciting the names of the family's (male) ancestors, and evokes the clan's totem (*dilizintaba*, 'mover of mountains'). Following this conventionalized, traditional performance, a second voice surfaces and expresses the writer's own pleasure at her own performance ('I get emotional when speaking our language'). The performance is punctuated by exclamations (*ahh, kwowu, tsii*), which remind the reader of the traditionally oral nature of such salutations.

(4) Ahh!Dilizintaba! Wangen'uMpemvu, uJali, uJuta, uNgciva, uTshantshane, uXhongwana uMaqath'alkhuni. Kwowu sizukulwana ndini sikaSam, sika-Moni, sikaFatuse kaGxoko.Tsii! Ndisukendizive ndinelunda madoda ngesintu sakowethu!!! (Facebook 2010)
'Ahh! Mover of mountains! Here comes Mpemvu, Jali, Juta, Ngciva, Tshanatshane, Xhongwana, the one with stiff ankles. Oh you descendants of Sam, of Moni, of Fatuse, of Gxoko. Tsii! I get emotional, people, when speaking our language!!!'

Jokes are a common example of small performances. Example (5) circulated as an SMS joke in Nigeria among university students. In the message an imagined writer is trying to produce a standard English greeting; however, with little success. The writer – in mock desperation – eventually continues in Nigerian Pidgin (underlined), mixed with Igbo (*chai* is an exclamation, bold) and Yorùbá (*jare* is an emphasis marker, italics). In this SMS joke we see the artful combination, and indeed struggle, of two voices: the aspirational voice of standard English, and the everyday voice of Pidgin.

(5) hello, how do U did? sory, how do U done? I man hw does u did? ah! devil is a liar. Hw u done? **Chai** na wah O! English hard O! Abeg hw u dey *jare*.[9] (SMS, Nigeria 2009)
'Hello, how do you did? Sorry, how do you done? I mean how does you did? Ah! The devil is a liar. How you done? **Oh**, it really amazes me! English is indeed difficult! Please tell me how you are, *really*.'

These small performances construct, through linguistic choices and the combination of forms, particular speaker personae. Example (3) projects a skillful, creative writer who also has a sense of humor; in example (4), the writer persona is skillful too as well as knowledgeable and appreciative of traditional African ways; the SMS in example (5) enacts – in a jocular vein – the voice of a struggling writer, desperate rather than eloquent. Ben Rampton (2009: 165) has suggested that the 'projection of a fairly well-delineated persona', that is, a social type, is a central aspect of speaking in a performance frame. This applies to both performance-2 and performance-3.

In the previous chapter I discussed Austin's notion of performatives, that is, utterances that can bring about a change in the social world. The chapter gave the example of an explicit performative (saying *I do* in a marriage ceremony). However, performatives are not limited to promises, requests or other explicit speech acts, and all language use needs to be understood as performative: 'to say something is in the full normal sense to do something' (Austin 1962: 94). What is the relationship between performatives and performance? Both nouns derive from the same verb ('to perform'), which has two meanings: (a) to carry out an action (roughly Austin's performatives), and (b) to present something to an audience. While performance foregrounds meaning (b), the noun performativity – which was introduced by Judith Butler (1993, 1997, 1999) in her work on gender – builds on Austin's notions of performatives as social action, that is, meaning (a). Both concepts – performance and performativity – have been influential in sociolinguistics, and they stand in a somewhat uneasy relationship to one another. It is clear that they are different and don't mean the same, but there is also a sense that they are 'neighboring concepts' and as such not entirely separate from one another (Velten 2012: 249; Schechner 2002). The remainder of this section discusses performativity and its relation to performance.

Performativity – which, like all language, depends on iterability and repetition for its success – is a fundamental feature of language-in-use: the moment we open our mouth (or send a text message), we *do things with words*; that is, we establish friendships and forge bonds, we break up with loved ones and reunite, we disagree and find solutions, we make others laugh and, sometimes, bore them. Moreover, through using language we also constitute ourselves as subjects in the world. Who we are, or who we wish to be, is not pre-discursively given (an identity we *possess*), but is produced in and through language and other sign systems (again it is something we *do*; see also Bucholtz and Hall 2004; Coupland 2009; Le Page and Tabouret-Keller 1985). Sometimes this doing has a certain theatrical quality to it – we might be performing to others a fairly well-established version of ourselves, or project an identity we desire (Goffman's performance-2). At other times, however, the process is emergent, and we become who we are not by strategically enacting a particular pre-defined persona, but through the way in which we engage in interaction. That is, certain ways of speaking/writing are perceived by others as indexical of social identities and ways of being. Virtual environments are of particular interest here because the absence of physical bodies, with all that they signify in interaction, means that other forms of signification – text and image – do most of the work.

Let us look at an example. In 2008, I conducted a chat experiment at the University of Cape Town with one of my research assistants, Oscar Sibabalwe Masinyana. We provided a group of five isiXhosa/English-speaking students with access to computers in a lab, and connected them to a specifically created public chat room (called #sharpsharp, drawing on the South African colloquialism *sharp*, which roughly means 'cool'). The only instruction participants received was 'log on, stay for two hours and see what happens'. By casting the interaction as an experiment, a performance frame was evident from the outset: the students knew that they were being recorded (and had agreed to this) and that I (as the researcher) would later analyze the transcript. Sitting in my office during the experiment, I decided to log on quickly. My idea was to be as unobtrusive and invisible as possible; I certainly did not want to impress an audience or present myself as a particular type of person. I simply wanted to lurk and to see how it was going. Example (6) is a shortened transcript of my six-minute stay in the room. My nick, engen, is the South African version of Shell, Caltex or BP, the name of a petroleum company.

(6) #sharpsharp, South Africa 2008 (isiXhosa in italics, Afrikaans underlined).

* engen (mori@sg-32065.uct.ac.za) has joined #sharpsharp

1	<@tokyo>	hi engen	
2	<engen>	*molo*	'hello' (singular)
3	<engen>	*molweni*	'hello' (plural)
4	<luvu>	*Yha*	'hi'
5	<@tokyo>	*molo* – who are u?	'hello'
6	<umgqusho>	HI engen.	
7	<Banda>	*molo nawe kunjani*	'hello to you too, how are you'

8	<engen>	hi hi	
9	<luvu>	Sharp engen joe	'cool'
10	<engen>	sharp sharp soos die naam van die praat kamer	'like the name of the chatroom'
11	<luvu>	wats up	
12	<luvu>	*uphi*	'where are you'
13	<Ziya>	*Hayi bo* why is everthing abt engen nw all of a sunday?	[exclamation]
14	<engen>	good point!	
15	<engen>	you are putting me on the spot:-(	
16	<@tokyo>	where u from engen? CT?	
17	<umgqusho>	i agree . . . engen, to the curb plz . . . next!;-)	
18	<Ziya>	*Wena* engen, wat's up with your Afrikaans/	'you'
19	<engen>	just felt like it	
20	<luvu>	come on petrol Yo engen	
21	<engen>	people always laugh about my name	
22	<Banda>	emagine if ur real name was Mngqusho, I would be hungrrr *qho xhandikubona*	'every time I see you' [Note: *umgqusho* is a local dish made of samp and beans]
23	<Banda>	Hahahah	
24	<@tokyo>	lol	
25	<engen>	hahah	
26	<luvu>	lol	
27	<engen>	ha	
28	<@tokyo>	banda *uyasibulala*!	'you are killing us!'
29	<engen>	uphi andie?	'where is andie?'
30	<Banda>	Ye engen	
31	<luvu>	*teta engen*	'talk engen'
32	<engen>	ok ok *ndiyahamba*	'I am leaving'
	engen (mori@sg-32065.uct.ac.za) has left #sharpsharp		

I returned to the room again briefly toward the end of the experiment, when I became the focus of umgqusho's virtual affection (discussed in Deumert 2014). I discovered only later – during the follow-up interviews with participants – that all participants had assumed me to be Black and male (while I am quite obviously White and female).

(7) Finding out that engen was a White woman was the cherry on top. That was so . . . brilliant. I was like 'what'? No way! Engen just was so masculine to me – BLACK!! And there she comes along . . . oh no, you did it! (2008, interview data, umgqusho)

What did I *do*? How did I manage to transform – quite unintentionally – from a White female into a Black male, using nothing but typed language? Whose voice did I articulate? Nicks are an important aspect of online interaction; they announce us as subjects and hail us into being. My choice of engen was meant to be neutral, indexing a neutral object, not a person. Yet objects – just like colors – can project gendered identities, and we can imagine an association along the lines of *petrol, cars, speed, men!* Similarly, I had hoped that by alternating between languages (Afrikaans, English and isiXhosa), I could remain unclassifiable in terms of ethnicity/race. Yet in South Africa, speaking an African language is indexical of being African. Historically, Africans learned English and Afrikaans, but non-Africans only rarely learned an African language. My performance – non-routine and crafted, but meant to be not special but ordinary and invisible – both succeeded and failed. I was able to take part in the conversation as an ordinary participant (and was not recognized as 'the professor'), but was assigned an identity I had not anticipated or desired. My interactional turns – minimal as they were – repeated a script that was read by others as indexing masculinity and Africanness.[10]

Can this example help us to clarify the relationship between performance – routine and non-routine – and performativity? In the preface to the second edition of *Gender Trouble*, Judith Butler argues for a close relationship between performance and performativity: 'the speech act is at once *performed* (and thus theatrical, presented to an audience, subject to interpretation) and *linguistic*, inducing a set of effects through its implied relation to linguistic convention' (1999: xxvii; my emphasis).[11] Richard Schechner (2002: 141) too emphasizes their close relationship when he notes that 'every social activity can be understood as a *showing* of *doing*' (my emphasis). That is, even ordinary actions such as greeting someone in the street or online can be *framed* as performances (as in example (2)). Although performance and performativity are not the same and have different disciplinary roots (theatre/art and speech act theory respectively), they overlap at times and interact in complex ways. Of the two, performativity is the broader, language-philosophical concept. Performativity is embedded in practices of repetition that produce meaning on the basis of the socio-historical associations of words and other linguistic forms. It is fundamentally a semiotic concept, which looks at how signs create – on the basis of their iterative and citational qualities – social meaning. It emphasizes the processural and constructed nature of social life, and, according to Butler, creates rather than presupposes the subject or the doer (example (6) see also Salih 2002). Performance is a narrower term and 'is something *the subject does*' (Kulick 2003: 140, my emphasis). That is, performance requires a doer and a certain degree of reflexivity. It is usually clearly marked and delimited (especially performance-3), as well as artful, that is, crafted and capable of evoking an aesthetic response from the audience (again, especially performance-3, but also, to a lesser degree, performance- 2). However, every performance relies necessarily on performativity, the citationality of signs that allows us to do things with language, for its effects (as illustrated in example (6)). In other words, all behavior is performative, but not all behavior is performance.

STYLIZATION: PERFORMING HETEROGLOSSIA

Not only identities are performed, a *doing* that is a *showing*, but languages, styles and varieties are performed too, and put 'on display' for an audience. Thus, the text in example (6) not only performs the persona of an uneducated speaker, but also puts West African Pidgin, Yorùbá and Igbo on display.

The notion of stylization, another key concept of Bakhtin ([1929/1963] 1984], is helpful in the analysis of such examples. Stylization can be distinguished from style, which describes the ways in which one acts habitually, persistently and routinely. The habitual nature of style is reflected in phrases such as 'this is my *style*', or when linguists refer to people talking as using an 'informal *style*'. Stylizations, on the other hand, are small performances in the sense of Bauman (performance-3). They are brief and fleeting moments in interaction when one speaks, quite self-consciously, with someone else's voice, when one 'puts on' a voice, displays a voice to an audience (Bakhtin [1934/1935] 1981: 362). The speaker thus moves beyond what is expected, and the 'disjunction of speaker and voice draws attention to the speaker herself/himself . . . positioning the recipient(s) as spectator(s), and at least momentarily reframing the talk as non-routine' (Rampton and Charalambous 2012: 484). Stylizations draw the audience's attention to the realities of heteroglossia, the presence of multiple social voices, and make diversity visible (Rampton 2009: 149).

In analyzing stylization, the notion of authenticity is important: am I putting on a voice that is entirely alien to me or a voice that is different from my ordinary conduct, but nevertheless part of who I am (or wish to be)? In example (6) there is no claim to authenticity. The joke is funny, rather than embarrassing, only because the represented persona – an unschooled writer who struggles with English – is far removed from the voice of the educated audience of university students. Examples (4) and (5), on the other hand, are performances of authenticity and the voice, which is adopted, is a desired voice as well as a voice to which the writer has legitimate access. Although English is used habitually by isiXhosa speakers in online contexts, in these two examples speakers show themselves to be in control of their language and culture, and display this knowledge for an audience. And the audience appreciates these displays. In response to example (4) another Facebook user commented admiringly: *wantsokotha Mhlekazi*, 'you are going complex [in terms of language], beautiful one [a traditional form of address]'.

A special case of stylization is what Ben Rampton (1995) has called crossing. In such instances the gap between the speakers'/writers' background and the voice they adopt raises not only 'the interactional question "why that now?"' but also 'the more political "by what right?"' (Rampton and Charalambous 2012: 485). My own episode as engen could have been interpreted as crossing if the others had known that I was a White woman, that is, if I had tried to sound Black. Yet since I managed to pass as authentic (a Black man), the question of entitlement and authenticity did not arise.

In spoken language, stylizations are often keyed by acoustic changes (pitch, voice quality, volume or speed of delivery). How can we identify stylizations in written interactions? Nikolas Coupland (2007: 1, 154) suggests looking for linguistic contrasts and violations of co-occurrence expectations as a methodological strategy: the unexpected ('why that now?') is a good indicator that something is happening. Writers can also draw on medium-specific ways of indexing style. For example, they could use fonts or colors to mark different voices or, in YouTube videos, utilize call-outs to visually insert a second, stylized voice. An example of this can be found in a YouTube video response to the controversial 'Kony2012' campaign, which was discussed at the beginning of Chapter 4. The clip was produced by a young Ugandan-American woman. In the audio stream she speaks with an American accent and there is no auditory trace of her Ugandan background. Using the call-out application, however, she inserts a second voice, which establishes her right to speak as a Ugandan. Despite her American accent she has not lost her cultural moorings and remains a *musoga*, a member of the *busogo* group from eastern Uganda (*oli otya* is a common greeting;).[12]

The examples discussed so far have addressed potentially global mass audiences, and even the chat room in example (6) was, in principle, open to anyone with

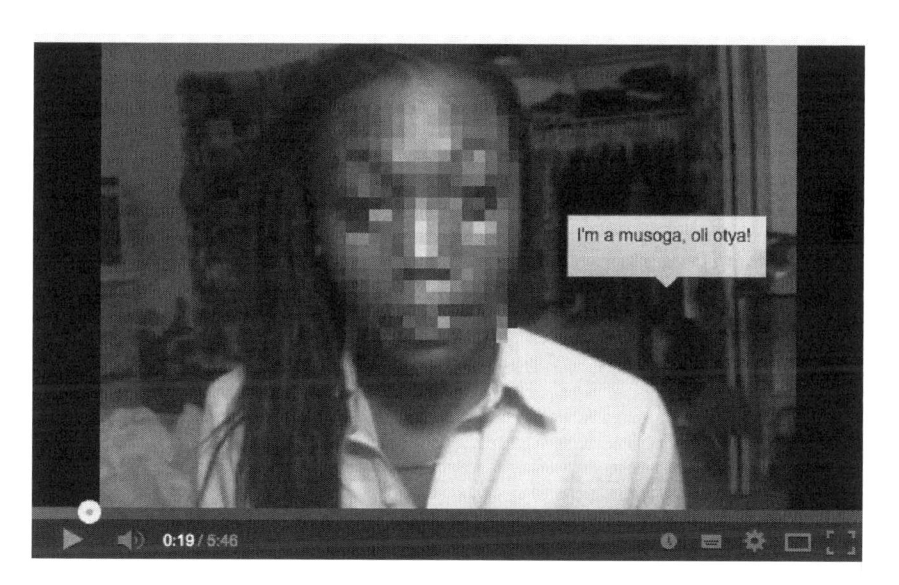

Figure 6.3 Call-outs (screenshot, August 29, 2013; http://www.youtube.com/watch?v=7DO73Ese25Y) – audio stream for segment accompanying the call-out: 'now I am 100% Ugandan, both my parents were born and raised in Uganda'

an internet connection. Example (8) is different: it is a private chat between two 18-year-old friends: @NN@ (female) and Old Money (male). Both come from a small South African town and Afrikaans is their main language of interaction. The habitual style of their online interactions is characterized by high-density non-standard spellings (further discussed in Chapter 7). Lexically it is peppered with colloquial vulgarities. In line 3, *kak leke* translates literally as 'shit nice', but means 'very nice' or 'really cool'. *Lmimp* and *lmj* are Afrikaans laughter acronyms that function similar to *lmao* ('laughing my ass off'). The Afrikaans acronyms represent overtly vulgar phrases with sexual connotations: *lmimp/lag my in my poes* translates as 'laughing myself in my cunt', and *lmj/lag my jas* is 'laughing myself horny' (for further discussion, see Chapter 8).

(8) Mxit data (Winter 2011)

		Mxit conversation	Standard orthography	Translation
1	Old Money	Jip hu was jo dag	Jip, hoe was jou dag	'Hey, how was your day?'
2	@NN@	Ngal oryt n joune?	Nogal alright en joune?	'Not too bad, and yours?'
3	Old Money	Nai kak leke	Nee, kak lekker	'Real cool'
4	@NN@	huso?	Hoeso?	'How so?'
5	Old Money	Nai my vrou was hie wt x miskien met gan trou	Nee, my vrouw was hier wat ek miskien met gaan trou	'Yeah, my woman was here, the one I might marry'
6	@NN@	Lmimp en wt van dai ene vn [name of a town]?	Lmimp en wat van die eene van [name of town]?	'*Lmimp* and what about the one in [name of a town]?'
7	Old Money	Nai sy is ng an my sy	Nee, sy is nog aan my sy	'No, she is still at my side'
8	@NN@	Lmj d van jule mans wt klmp myre mt ht. Ai ai ai	Lmj die van julle mans wat klomp myre moet het. Ai Ai Ai	'*Lmj*, those of you men who must have many wenches. Ai ai ai'
9	Old Money	Lmj	Lmj	'*Lmj*'
10	@NN@	G2g mwah	Got to go mwah	'Have to go, mwah'

The non-standard orthography and the heavily colloquial lexicon locate this interaction outside of what one might call 'good manners', and the transgressive-subversive mood is also evident in Old Money's display of hyper-masculinity, boasting about his two girlfriends and showing himself to be a ladies' man. In her response @NN@ adopts, just for *a moment*, a different voice (marked in bold, line 8). This is a historically distant voice and contrasts starkly with the rest of the short interaction. The noun *myre* (a dialect variant of *meide*, 'servant girl') is archaic and unlikely to be used in conversation by anyone younger than sixty. Moreover, its connotations are negative, linking back to the history of apartheid and the exploitation of local maids in White, colonial households. It is followed by a lament-like exclamation (*ai ai ai*), indexing the resigned voice of age, a gentle reprimand of Old Money's promiscuity: '*Lmj*, those of you guys who must have many wenches. Ai ai ai.' Here stylization is not about enacting a well-defined social persona, but links to performativity as discussed in the previous section. @NN@'s comment constitutes Old Money as a particular kind of person; it is an interpellation to which he responds with laughter. Thus, while stylizations can be performed to *impress* an audience, they also allow us to *express* a stance, that is, they can be used as a form of evaluative language (Rampton 2009; Snell 2010).

To conceptualize speaking/writing not as the instantiation or animation of a structured, pre-existing system, but as the creative combination of heteroglossic resources, also reconceptualizes the notion of linguistic competence. Individual words, such as *myre*, can be inserted even if the speaker/writer is not actually proficient in the register or dialect to which the word belongs. And, sometimes, successful stylizations can be made even if one does not know a single word of the intended language. The brief SMS interaction in example (9) took place between two male, English-dominant South African students in their twenties as they were making plans on a Friday evening (Cape Town, 2010; Afrikaans in bold, isiXhosa in italics).

(9) Dustin: *Molloboetie*, **hoe gaan dit**? What's the plans for tonight? ('Hello brother, how are you?') [The correct spelling for the isiXhosa greeting would be 'Molo boetie'.]
Carlo: Shabash shabash, gildie gildie. What do you say we hit jade? [Jade is a local night club.]

Carlo reflects on the interaction as follows:

(10) In the first message sent, Dustin used three different languages, having gone to a boarding school where learning three languages was possible. He uses three languages in one sentence *to display* his level of intelligence as well as practising his writing in all three languages. The message I sent to Dustin was *to display* my form of multilingualism and intelligence by saying something in Arabic. Even though what I said was not a greeting, it still made me come across as more intelligent. (My emphasis)

Carlo describes Dustin's use of Afrikaans, English and isiXhosa as a deliberate 'display', or performance, of linguistic skill (notwithstanding his misspelling of the isiXhosa greeting) and feels under pressure to show himself as multilingual too. His response is to write in what he calls 'Arabic'. Having some knowledge of Arabic is not unusual in Cape Town, which has a sizable Muslim minority. However, what

Dustin wrote is not Arabic at all: *shabash* is Urdu, meaning 'well done' (possibly picked up from watching the Pakistan cricket team); *gildie* is dubious – it might be purely fictional or could be a misspelling of *jaldi*, 'quick', which is Urdu as well. The phrase is artful in its use of reduplication, and is meaningful as a display of multilingualism. It resembles what Jan Blommaert (2012a) has called 'lookalike language' and what Claire Kramsch (2009: 11) describes as 'myth' in her work on the *Multilingual Subject*: an example of language that 'focuses on the aesthetic, that is, perceptual aspects of words and on the affective impact of their connotations'. What matters here is that it is perceived as a display of linguistic skill, and the writer shows himself to be heteroglossic, not monologic and monolingual. His utterance works performatively within the context where it is enacted, even though it is, from a strictly linguistic perspective, a 'failed' performance of 'Arabic'.

CONCLUSION: A SHARPLY HETEROGLOT ERA?

Bakhtin suggests in 'Discourse in the Novel' that although heteroglossia is always present, lurking in the background even in the most centripetal societies, some historical periods are more heteroglossic than others. Those are times 'when the collision and interaction of languages is especially powerful, when heteroglossia washes over literary language from all sides [and] . . . aspects of heteroglossia are canonized with great ease' ([1934/1935] 1981: 418). Contemporary social life – referred to as late modernity or liquid modernity – is characterized by the increasing mobility of people and resources as well as an emphasis on choice and self-invention (Giddens 2011; Bauman 2012; Archer 2012; Elliot 2013). This provides fertile ground for the play of centrifugal forces; that is, for the creative, rather than predictable, use of language and other semiotic resources in fashioning and performing identities, thus showcasing the diversity that surrounds us.

Intertextuality (as discussed in the previous chapter) and heteroglossia (as discussed in this chapter) point to the same phenomenon; that is, the fact that every text is essentially 'a mosaic of quotations' (Kristeva [1969] 1980: 66), and that meaning resides in our ability to cite and combine forms that existed prior to the here and now. However, there are also differences between intertextuality and heteroglossia. Firstly, work on intertextuality typically focuses on texts and by extension linguistic structures, not on speakers/writers. This 'anti-human' position is famously summarized in Roland Barthes' essay 'The Death of the Author' ([1967] 1977), which uncoupled text and author. Bakhtin, on the other hand, does not seek to announce the author's epistemological 'death' and speakers/writers remain central to his thinking. Secondly, heteroglossia, unlike intertextuality, which usually focuses on texts as combinations of signs, encourages us to shift our gaze – momentarily – to language *below* the sentence and toward the indexicalities of individual linguistic forms. This does not mean that we want to move away from language *beyond* the sentence. 'Momentarily' is important here: it is the back and forth, from individual forms to their co-articulation within an utterance, that defines this approach. Bakhtin's work opens up a way of looking at language – below and beyond the sentence – which can combine the linguist's rigor and skill in the description of forms with a study of meaning that recognizes language's inherent multi-vocality.

The fundamentally heteroglossic nature of language is essential to creativity. Heteroglossia provides speakers and writers with resources, whether multilingual or multidialectal, for expressing stances, identifications and social personae. Such practices are particularly visible and marked in moments of 'heightened attention' (artful performances and linguistic stylizations of social voices), but also shape the broader dramaturgy (Goffman's performance-2) and general performativity of social life. Reflexivity is an important aspect of both performance-2 and performance-3 – routine and non-routine crafted displays – and is facilitated by the affordances of writing. The next chapter looks at another affordance of writing that is important in the creation of written verbal art: its visuality.

NOTES

1. Bakhtin's work has inspired sociolinguistic thinking since it became available in English translation in the early 1980s (e.g. Hill 1995; Woolard 2004; Blackledge and Creese 2010; Bailey 2012), and has also been influential in new media studies (e.g. Jones and Schieffelin 2009; Androutsopoulos 2011; Leppänen and Peuronen 2013).
2. Speech is, of course, chronologically prior to writing, which only dates back around six thousand years (whereas speech is probably as old as *Homo sapiens*).
3. Even though twentieth-century linguistics has prioritized spoken language, it has generally studied it through writing, i.e. transcripts.
4. Research has largely focused on English and other European languages, i.e. languages that have a well-established written norm. These observations cannot be applied to written language in general (e.g. Besnier 1995).
5. However, there are obviously areas within sociolinguistics, such as the study of sound change and articulatory phonetics, where spoken language data remains primary.
6. All data – no matter how large the set – is specific to a particular context and one needs to be careful with generalizations. Thus, while Tagliamonte and Denis found high levels of variability in their Canadian IM data, Jones and Schieffelin (2009), looking at a sample of university students in New York, found *be like* to be an emerging norm in their 2006 IM data (but not in their 2003 data).
7. This is not the place to discuss the debates about authorship surrounding Bakhtin's oeuvre (for an overview see Vice 1997).
8. Bakhtin's centripetal/centrifugal distinction is echoed by Robert Le Page's discussion of focused vs. diffuse language situations (Le Page and Tabouret-Keller 1985).
9. /O/ is an emphasis marker in Nigerian Pidgin; 'the devil is a liar' is a common Nigerian exclamation.
10. This example brings to mind Don Kulick's (2003) discussion of a heterosexual male uttering 'no' in response to a sexual invitation from a woman. Irrespective of the speaker's intentions, uttering 'no' can quite involuntarily signify the speaker as homosexual. Similar processes were also at work in the audience responses to 'Xhosa lesson 2', discussed in the previous chapter.
11. In earlier publications, however, Butler had rejected theatrical notions of performance (for example, in *Bodies that Matter*, 1993).
12. The creator of the video has not been active on YouTube since 2012 and I was not able to contact her for permission to reproduce a screenshot with her image. The video itself is, however, in the public domain.

Chapter 7

Textpl@y as poetic language

> New game inspired by someone asking what YOLO meant & being told 'you obviously love owls' . . . alternative textspeak meanings! ROFL, LOL, etc.
>
> Twitter 201

INTRODUCTION: WRITING ✍ IS LIKE A PICTURE

In 1946, Dwight Bolinger published a paper about the visual side of language. One of the examples he discusses is: *he ghulped and ghasped.* Rendering <g> as <gh> creates associations with those words where the velar plosive [g] is represented as <gh>: *ghoulish, ghost, ghastly* and *aghast. To gasp* is not the same as *to ghasp* – the <h> suggests breathiness, evokes an eerie feeling of supernatural doom. More recently, Gunther Kress (2000, 2010) and Jan Blommaert (2008) have emphasized the visuality of writing, the ways in which writing is *always* 'a kind of picture' (Kress 2000: 52). The link between writing and drawing is not limited to digital texts, but also common in the early writings produced by children and well documented in history. Medieval manuscripts, for example, were heavily visual, and handwritten vernacular texts, such as cookbooks or letters, paid due attention to the aesthetics of writing (Basso 1974). The visual qualities of writing did not disappear when typing took over from handwriting: typing allows choice of fonts and colors, and icons can be inserted (such as ✍ above).

The way in which the visual nature of writing enters into meaning-making becomes evident if one attempts to read out texts that play around with spelling, punctuation and the arrangement of words. Consider the following example, written by Enrico, a South African writer in his twenties:

(1) Girl. U. So. Booty shakin. .Heart breakin..Steamy hot. Neva.Stop..Short. Skirt. Love 2 flirt. .Angel babz. Spoild. .Maybe. .Tight jeanz nd curvy hipz. .High class . Ghetto ass. .. .Sexi smile. .Blazin. Style. .Lushious .Thighs ..Kandy eyes. .Killa kiss .Sexy. Thong nd glossed lipz. .O.M.W baby. . . Hw. Cud i resist ! (Facebook status update, 2009)

How should one render the unusual punctuation of *.Girl. U. So .Booty shakin* in spoken language? And what about *Kandy eyes*? Is *kandy* the same as *candy* and can thus be read out simply as [ˈkændi]?

In popular discourse, the type of writing illustrated in example (1) has generally been evaluated negatively: although hyperbolic fears of a full-blown onslaught on the education system show signs of cooling down, there remains a feeling that non-standard, digital writing is not desirable and should not be encouraged; it is believed to damage reading and writing abilities, thus, ultimately, threatening the survival of literacy norms (for examples see Thurlow 2006). The image in Figure 7.1 captures the lingering fear that we are witnessing the decline, if not loss, of a 'superior' and 'better' literate culture where writing was enshrined in books and libraries; a time when writers followed orthographic norms diligently and respectfully.

Yet, contrary to such popular and populist fears, research indicates that non-standard writing – far from being a scourge – has educational benefits and might help, rather than hinder, formal literacy by developing metalinguistic awareness (Plester et al. 2008; Wood et al. 2011). This chapter approaches digital non-standard writing from the perspective of creativity and aesthetics, as an impressive

Figure 7.1 *A dictionary commits suicide . . . every time you write using the letter K in place of the Q* (circulated on Facebook in 2013). In Spanish and Italian texting, *k* can replace *c* (e.g. *kasa* 'house'), and can stand for the complementizer *que* [ke] 'that' (Back and Zepeda 2013)

display and performance of linguistic skill, and as an everyday example of artful, poetic language (see also Thurlow 2012). The focus of the chapter is on the visual side of language, that is, typography, the shape and arrangements of letters (literally 'form writing'), and orthography, sets of conventions of how to spell certain words (literally 'correct writing').[1]

POETIC LANGUAGE: FOCUS ON FORM

Interest in poetic, artful language has a long tradition. An early example is Aristotle's *Poetics* (335 BC), which located the poetic in 'everything that diverges from standard use' and favored 'alien expressions', that is, dialectal or obsolete words. And throughout the centuries there have been reflections on what makes a text poetic by poets and literary scholars (Silk 2010). In linguistics the notion of poetic language is strongly associated with the work of the Russian linguist Roman Jakobson. Jakobson (1960) approaches language as fundamentally multifunctional; that is, utterances *always* do more than just one thing in communicative encounters. Language not only allows us to talk about the world (referential function); it also expresses how we feel (emotive function) and allows us to address others (conative function) and to establish social connections (phatic function). Imagine I were to utter the following phrase: *what a beautilicious, sun-kissed day my dear!* In doing so I would not only describe the weather (referential function), but also express my delight (emotive function) and my orientation toward an addressee with whom an affectionate bond is established (conative and phatic function). But there is more to language than talking about the world to others: sometimes language itself becomes the object of our attention. The neologism *beautilicious* and the metaphor *sun-kissed* illustrate what Jakobson calls the poetic function: the focus is on linguistic *form*, the manipulation of the shape and meaning of words. And my willingness to play around with language suggests a particular attitude toward language, namely, that it is appropriate for speakers to be creative and unconventional. This would be an example of the metalinguistic function (which could, of course, also go in the opposite direction and display an attitude that values correctness and adherence to the standard).

Jakobson argued that poetic language with its focus on form allows us to perceive 'the palpability of signs', and to experience 'the fundamental dichotomy between signs and objects' (1960: 356). What does this mean? When the referential function is foregrounded in communication – for example, when we are asking someone for directions ('Where is the post office?') – we treat the word ('post office', signifier) as a direct representative of the object we are looking for (the physical place we want to go to, signified). We do not ponder the fact that language works through mediation and representation, and that words are separate from the objects we are trying to locate. When the poetic function is dominant, however, the signifier is foregrounded; the shape and sound of words, the syntactic patterning of sentences and – in the case of written language – their visual representation. Words are not merely perceived as signs that represent objects in the world, but are themselves seen as objects that can be manipulated so that new meanings can emerge, and new associations can be established.

Jakobson's attention to the poetic might come as a surprise to those who see him as the godfather of linguistic structuralism.[2] Yet the poetic is central to Jakobson's theory of language. And like Sapir (Chapter 2), he was a published poet, and deeply involved with the Russian Futurist movement of the early twentieth century. Futurism shaped not only Jakobson's poetry but also his thinking about language, and it is therefore important to sketch the broad contours of this avant-garde movement.

Futurism – the name itself a programmatic celebration of the future, of modernity and technology, a departure from the past and tradition – was an artistic movement that involved writers, painters and performance artists. It was fundamentally anti-establishment and intentionally provocative in its aesthetic practice (for overviews see Perloff 1986; Lawton and Eagle 2004). Writers played around with language. They created untranslatable words, phonemic puzzles and onomatopoetic sound sequences; they used typography in non-conventional ways and manipulated spelling norms. At a time when linguists like Ferdinand de Saussure ([1916] 2013) described writing as just an imperfect transcription of speech (Chapter 6), Futurist poets saw writing as a form of painting, full of creative and expressive potential. They called – in metaphorical language – for the 'liberation' of not only the word, but also the letter from the drabness of convention. Letters should be dressed beautifully and not all 'wear the same government overcoats . . . lined up in a row, humiliated, with cropped hair, and all equally colourless, gray' (Kruchenykh and Khlebnikov 1913, in Lawton and Eagle 2004). In his poem *Akhmet* (1912), Aleksei Kruchenykh put this programmatic statement into practice. He combined capitals and small letters, as well as typeface and handwriting (for a reproduction of the poem see Perloff 1986: 140). Instead of dressing the words he uses in 'the same government overcoats', he creates unusual word-shapes, such as ГЕИеРаЛ, *GENeRaL*. The aim of such practices was 'to make strange' (остранение, *ostranenie* 'estrangement'); that is, to disrupt the habits of reading and the semiotic patterns of everyday communication; to de-automatize and defamiliarize perception. Rather than looking 'through' the word toward the referent, the reader (or listener) would now 'have to pause and look at it' (Sturrock 1993: 101).

Futurism, its internal fragmentation notwithstanding, was a global artistic movement that influenced artists in Paris, Moscow, Rome, Rotterdam, Berlin, Lisbon, London, Madrid, Rio, Tokyo and New York. It included Dada (in Germany), Cubism and Surrealism (in France and Spain), and Vorticism (in England). Figure 7.2 reproduces Guillaume Apollinaire's (1880–1918) visual poem *Il pleut* ('It is raining'). The arrangement of letters foregrounds the visual (representing falling rain), and 'makes strange' the written message, which is arranged from top to bottom rather than left to right.

The poem belongs to the collection *Calligrammes* (1918).[3] The title is a programmatic neologism combining 'calligraphy', that is, 'beautiful writing', and 'telegram'. The notion of the telegram is important here. It refers not merely to

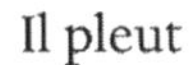

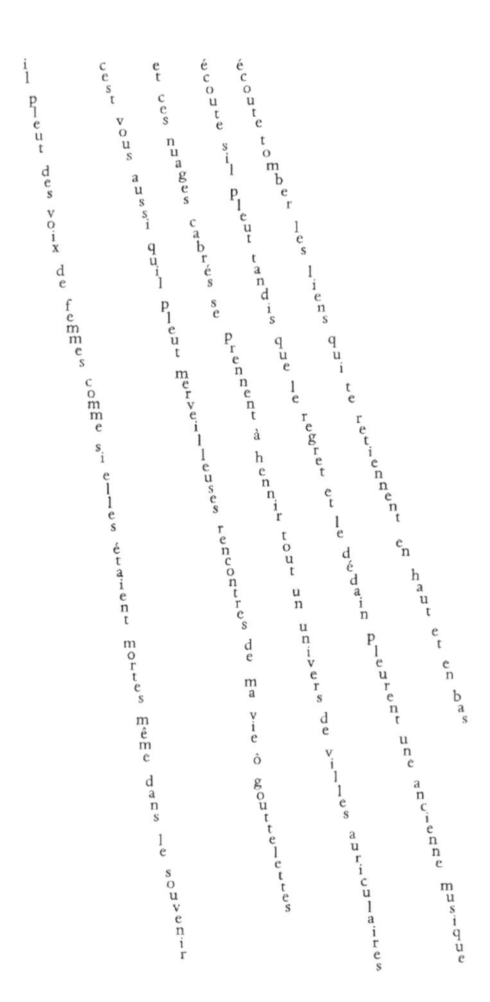

Figure 7.2 Guillaume Apollinaire's poem *Il pleut* ('It is raining', 1918). Translation: 'It's raining women's voices as if they were dead even in memory / It's raining you as too marvellous encounters of my life oh droplets / And those clouds rear and begin to whinny a universe of auricular cities / Listen to it rain while regret and disdain weep an ancient music / Listen to the fetters falling that bind you high and low'.

a new technological mode of transmission but also to the aesthetics of producing short and condensed texts, giving expression to what the Italian Futurist Filippo Marinetti ([1909] 2009) called 'the beauty of speed'. Julia Gillen (2013) notes, in her work on the Edwardian picture postcard, that the early twentieth century shared many of our current concerns: a sense of living in a culture of speed, shaped by ever-changing technologies, globalization and increased mobility. New communication technologies, which transmitted information quickly and emphasized brevity of expression, became the symbol of these changes. This includes the telegram, and also the telephone and the postcard. Since the early 1990s, this aesthetic has been remediated in the form of text messages as well as Twitter and Facebook status updates (on remediation see Chapter 1).

The literary scholar Majorie Perloff (1986) has argued that Futurism has cast 'a long shadow', and since the 1960s, especially, there has been a revival of many Futurist art forms, including collage, manifesto and impromptu performance art as well as visual poetry. Experiments with language – spoken and written – are common, and the boundary between high art and popular culture has blurred. The manipulation of orthographic and typographic conventions became common practice in sci-fi, graffiti, advertising, comics, fanzines and pop music (Kataoka 1997; Androutsopoulos 2000; Sebba 2007; Alexander 1929 and Pound 1923 and 1925 show the time depth of such practices in advertising). Pop music especially supported the global spread of many of these forms and linked them closely to English as *the* idiom of popular culture. In the late 1970s, bands with names such as XTC, U2 and INXS reached global fame, and from the early 1980s onwards the artist Prince used forms such as *u* (for 'you') or *2* (for 'to(o)') in his published song lyrics. He took the poetic, visual play to yet another level when, in the 1990s, he changed his name to a deliberately unpronounceable symbol (Figure 7.3; he has since gone back to 'Prince').

Figure 7.3 The artist formerly known as Prince, or the love symbol. The symbol has been explained as a combination of ♂ and ♀, representing 'male' and 'female'

Whenever I discuss popular culture examples of creative typography with friends and colleagues, many spontaneously volunteer anecdotes about seeing and using them in everyday writing long before mobile phones or widespread personal access to computers. Forms also spread globally. Deborah Spitulnik (1996) mentions a handwritten letter sent to her by a 14-year-old Zambian girl in late 1980s. At the end of the letter, the girl writes *over '2' you* – displaying 'conversancy with the latest trends in popular expression' (even though the respelling *2* was still felt to be somewhat unusual and remains in quotation marks).

The digital expressive space has recontextualized such practices further. The image in Figure 7.4 uses available ASCII symbols to create a virtual birthday cake. Figure 7.5 is the first of a series of onomatopoetic and deliberately ambiguous texts that were posted in the Facebook group 'Jokes in Xhosa'. These everyday examples are not unlike the visual poetry of the Futurists.

While Figures 7.4 and 7.5 are stand-alone visual texts, smaller typed images can be integrated into interactive, conversational writing. The best-known examples of this are emoticons, which come in different forms, shapes and styles, and are fairly conventionalized (Dresner and Herring 2012). In addition, writers create images spontaneously using ASCII symbols. Multimodality, in other words, does not require access to high-end computing (video and audio), and any keypad or keyboard can be used to produce multimodal texts. Examples (2) and (3) come from Marisol Del-Teso-Craviotto's (2008: 259) work on America Online (AOL) dating chat rooms (the first example from a heterosexual room, the second from a lesbian room). The information is presented linguistically ('coffee and roses', 'tits') as well as in the form of a picture.

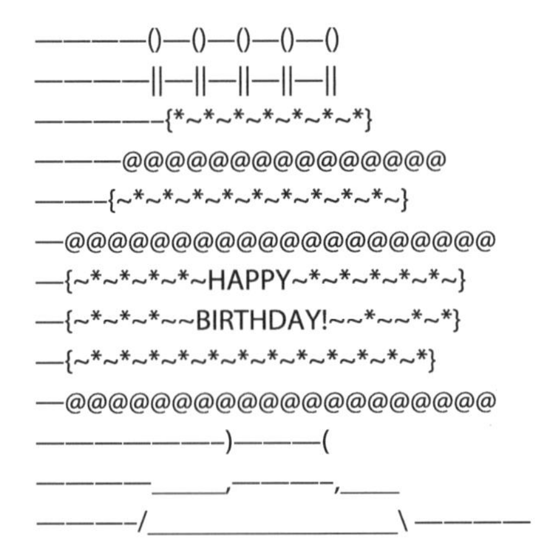

Figure 7.4 A textual birthday cake (Facebook 2011, on the occasion of an eighteenth birthday, South Africa; female English–isiXhosa bilingual writer)

ukhe uzive iinyosi?? Zzzzzzzzzzzzzzzzzzzzzzzzzzzzzz
zzzzzzzzzzzzzzzzzzzzzzzzzzzzzzzzzzzzzzzzzzzzzzzzzzzzzzzzzzzzzzzzzzzzzzzzzzz
zzzzzzzzzzzzzzzzzzzzzzzzzzzzzzzzzzzzzzzzzzzzzzzzzzzzzzzzzzzzzzzzzzzzzzzzzzz
zzzzzzzzzzzzzzzzzzzzzzzzzzzzzzzzzzzzzzzzzzzzzzzzzzzzzzzzzzzzzzzzzzzzzzzzzzz
zzzzzzzzzzzzzzzzzzzzzzzzzzzzzzzzzzzzzzzzzzzzzzzzzzzzzzzzzzzzzzzzzzzzzzzzzzz
zzzzzzzzzzzzzzzzzzzzzzzzzzzzzzzzzzzzzzzzzzzzzzzzzzzzzzzzzzzzzzzzzzzzzzzzzzz
zzzzzzzzzzzzzzzzzzzzzzzzzzzzzzzzzzzzzzzzzzzzzzzzzzzzzzzzzzzzzzzzzzzzzzzzzzz
zzzzzzzzzzzzzzzzzzzzzzzzzzzzzzzzzzzzzzzzzzzzzzzzzzzzzzzzzzzzzzzzzzzzzzzzzzz
zzzzzzzzzzzzzzzzzzzzzzzzzzzzzzzzzzzzzzzzzzzzzzzzzzzzzzzzzzzzzzzzzzzzzzzzzzz
zzzzzzzzzzzzzzzzzzzzzzzzzzzzzzzzzzzzzzzzzzzzzzzzzzzzzzzzzzzzzzzzzzzzzzzzzzz
zzzzzzzzzzzzzzzzzzzzzzzzzzzzzzzzzzzzzzzzzzzzzzzzzzzzzzzzzzzzzzzzzzzzzzzzzzz
zzzzzzzzzzzzzzzzzzzzzzzzzzzzzzzzzzzzzzzzzzzzzzzzzzzzzzzzzzzzzzzzzzzzzzzzzzz
zzzzzzzzzzzzzzzzzzzzzzzzzzzzzzzzzzzzzzzzzzzzzzzzzzzzzzzzzzzzzzzzzzzzzzzzzzz
zzzzzzzzzzzzzzzzzzzzzzzzzzzzzzzzzzzzzzzzzzzzzzzzzzzzzzzzzzzzzzzzzzzzzzzzzzz
zzzzzzzzzzzzzzzzzzzzzzzzzzzzzzzzzzzzzzzzzzzzzzzzzzzzzzzzzzzzzzzzzzzzzzzzzzz
zzzzzzzzzzzzzzzzzzzzzzzzzzzzzzzzzzzzzzzzzzzzzzzzzzzzzzzzzzzzzzzzzzzzzzzzzzz
zzzzzzzzzzzzzzzzzzzzzzzzzzzzzzzzzzzzzzzzzzzzzzzzzzzzzzzzzzzzzzzzzzzzzzzzzzz
zzzzzzzzzzzzzzzzzzzzzzzzzzzzzzzzzzzzzzzzzzzzzzzzzzzzzzzzzzzzzzzzzzzzzzzzzzz
zzzzzzzzzzzzzzzzzzzzzzzzzzzzzzzzzzzzzzzzzzzzzzzzzzzzzzzzzzzzzzzzzzzzzzzzzzz
zzzzzzzzzzzzzzzzzzzzzzzzzzzzzzzzzzzzzzzzzzzzzzzzzzzzzzzzzzzzzzzzzzzzzzzzzzz
zzzzzzzzzzzzzzzzzzzzzzzzzzzzzzzzzzzzzzzzzzzzzzzzzzzzzzzzzzzzzzzzzzzzzzzzzzz
zzzz

Figure 7.5 *Ukhe uzive iinyosi?* 'Do you sometimes hear the bees?' (Facebook, posted on 'Jokes in Xhosa', 2013; male isiXhosa–English bilingual writer in his twenties)

(2) TNCharmer: c(_) @}}--- coffee and roses for the ladies and hello room

(3) HOTAZHEAT88: COME HERE I GOT TITS (@) (@)
Xboigrlx2227: OOOOOO NICE AND PERKY AND FIRM
Da1nonlyteas: my good look at the size of Hot nips

Online communication is saturated with typographic play (Danet 2001; Goriunova 2012). The text in example (4) belongs to the notorious genre of emotionally manipulative chain letters. Unless the letter is passed on, something terrible will happen to the receiver. This particular letter has circulated on the internet at least since 2007 and remains popular. Again, the unusual typography enters into the process of meaning-making by 'making strange' and thus unleashing further meanings. A 'littlɛ Dɛad giяl' – eerie and scary in its unfamiliarity – will appear. A 'little dead girl', written in standard orthography, seems rather quaint by comparison.

(4) Tʜis isnt fakɛ appaяɛntlч if u copч pastɛ tʜis to tɛn pɛoplɛ in tʜɛ nɛxt tɛn minutɛs u will ʜavɛ tʜɛ Bɛst daч of uя lifɛ tomoяяow u will ɛitʜɛя gɛt a kiss oя askɛd out if u Bяɛak tʜis cʜain u will sɛɛ a littlɛ Dɛadgiяl in uя яoom TONIGʜT In 53 mins sumonɛ will saч i ovɛ u im Soяяч oя i wanna go out witʜ чou

Following Jakobson, the focus of this chapter is on *form*, that is, on the *linguistic products* of visual, written creativity. Non-standard spellings and typographic play are no longer the sign of the artist, but have become a hallmark of everyday digital writing. They are hated by some, cherished by others, but impossible to ignore.

'YOUR MYSPACE NAME *MUST* CONTAIN SYMBOLS AND INCORRECT SPELLINGS'

The title of this section comes from an old-ish (2005) online manual that advised writers on 'How to be cool on MySpace' (boyd 2008: 129). Although MySpace has since been overtaken by Facebook, the imperative to deviate from the normative standard remains an important maxim of informal digital writing. Those who use them construct non-standard spellings as 'markers of cool'. These index insider status and allow writers to 'dramatize', that is, perform a fun-loving and creative persona (boyd 2008; performance-2 as discussed in the previous chapter, i.e. in the routine sense of Goffman).

This does not only apply to college students in the United States. In her ethnographic work on texting in Kenya, Janet McIntosh (2010) has argued that English non-standard spellings afford writers 'a means of "showing-off" that one is "modern" . . . "developed", "fashionable", "Western", "dot-com", or a "town boy"'. In other words, it is not only access to new technologies that marks people as 'modern', but also the way in which they use them. McIntosh refers to such spellings as 'a global English medialect'. The term 'medialect' is useful as it draws attention to the mediatized nature of these forms, and their often conventionalized nature. However, it is also misleading as it suggests that we are dealing with a lect, that is, a variety that has fairly well-defined structures and norms. Rather, I would argue, we are looking at *strategies* for writing, not specific forms that are imitated and copied. These create recognizable but diffuse and heteroglossic 'media idioms' rather than well-defined lects (on media idioms, see Jacquemet 2005).

In order to describe the strategies that allow writers to create media idioms, it is useful to briefly consider what options we have at our disposal when representing language visually. Linguists distinguish between logographic and phonographic writing systems. In logographic systems, signs do not carry phonetic value, but instead refer to word-concepts, that is, morphemes. Writing is thus linked not to sound, but to meaning. Numerals are an example: the sign 4 represents the number 'four', but not the sound sequence [fɔ:(r)] in English, or [fi:ɐ̯] in German. In

Table 7.1 Writing strategies used in informal digital writing

Category	Examples (for English)	Writing system
Pictograms	♥ 'love', ☺ 'smile'	Logographic
Logograms	*xoxo* 'hugs and kisses', @ 'at'	
Acronyms	*LOL 'laughing out loud', lib 'lying in bed'*	Mixed system with logographic and phonographic elements
Rebus writing	*b4* 'before', *bcoz* 'because', *aQr8* 'accurate', *xams* 'exams'	
Consonant writing	*Txtng* 'texting', *swdrms* 'sweet dreams'	
Phonetic spellings	*guyz* 'guys', *dat* 'that', *goin* 'going', *hav* 'have'	Phonographic
Aesthetic writing	*@t h0m€* 'at home', *hugsz* 'hugs'	

phonographic writing, signs represent sounds. Phonographic writing is further separated into syllabic and alphabetic; that is, a symbol stands either for a syllable (typically a consonant–vowel combination) or for individual sound segments. Italian, for example, is written using an alphabetic system, Chinese languages use a logographic system and Cherokee a syllabic system. Digital writers combine logographic and phonographic writing in creative and often complex ways. The writing strategies illustrated in Table 7.1 are not only common in English, but are applied cross-linguistically. The following discussion illustrates their use in a variety of languages.[4]

Pictograms, such as ♥ or emoticons, are an example of logography. However, unlike conventionalized logograms, they work according to the principle of iconicity: there exists a relationship of resemblance between sign and meaning. Forms such as @ 'at', on the other hand, are conventionalized. Acronyms, such as *LOL*, are logographic if they stand for entire phrases ('laughing out loud'). If, however, they are read out phonetically as syllabic words or as a letter sequence (*lol* or *L-O-L*), then they follow the phonographic principle. Acronyms are common in digital writing cross-linguistically, and a particular enjoyment appears to be their lack of transparency. Unless one is an insider and knowledgeable, the letter sequence itself remains obscure. In German, for example, we find forms such as *hdgdl, hab' dich ganz doll lieb*, 'love you very much', and isiXhosa speakers have playfully localized *ROFL*, 'rolling on the floor laughing' as *GPY* (*giligidi, phantsi yintsini*, 'falling down laughing'), and *GBPY* (*giligidi bhu phantsi yintsini*, 'falling down laughing with a bang').[5]

Rebus writing has a long history: it was popular in ancient Greece and Rome, during the Middle Ages and the Renaissance and in Victorian times, and continues to capture our attention today. Rebuses are linguistic puzzles in which pictograms, numbers and letters are interpreted as syllabograms. Thus, a particular shape, such as *4* or *c*, no longer expresses a certain meaning – the numeral four or the third letter of the Latin alphabet – but a sound sequence, [fɔ:(r)] and [si:] respectively. *4* can thus be used to write the preposition 'for' and *c* can stand for the verb 'see' or the noun 'sea' (*C food*). They can also represent syllables within words, as in *be4* or *4get*. More complex rebuses work across word boundaries; for example, *ICQ* 'I seek you' in English or *W817* in Dutch. The latter constitutes the sequence 'w-acht-één-zeven' ('w-eight-one-seven'), which, when read aloud, sounds like the phrase *wacht eens even*, 'wait a minute' (Blommaert 2012b: 10).

Rebus writing has also been described for German (Bieswanger 2006), Italian (Herring and Zelenkauskaite 2009) and French (Anis 2007), as well as the Chinese languages. In example (5), Nek, who is in his mid-twenties and works in Taiwan as an IT professional, signs out of his IM chat with 'good-bye' (Yang 2009). It is spelled using Chinese characters as *gu-de-bai*. The Chinese word meaning of the three characters ('to murmur', 'virtue', 'to worship') is unrelated to the intended meaning. Only a bilingual reader, who knows both the sound values of the Chinese signs and the English meaning of these sound values, will be able to decipher this phrase.

(5) 咕 德 拜
gū dé bài
to murmur virtue to worship

Multilingual writers can also assign two (or more) sound values to the same written symbol. In Kenya, numerals carry both the English and the Kiswahili sound value. Thus, in example (6) the numeral *1* stands for English [wən] as well as Kiswahili *moja* [mɔʤɑ]. The numeral 2, however, only carries its English value in this example.

(6) 2dy @ 8 sm1 shud b thea 2 c u b4 u go . . . 2ko pa1;-)
'today at eight someone should be there to see you before you go . . . tuko pamoja' ('we are together'; Barasa 2010: 130)

Jan Blommaert (2012b: 10) describes similar practices for Dutch. In example (7) the Dutch phonetic value of the numeral 3 is inserted into an otherwise English sentence, producing what he calls 'English with an Antwerp accent'. Nadia Lawes gives examples of bilingual rebus writing for Loniu, an Austronesian language spoken by fewer than five hundred people in Papua New Guinea (Temple et al. 2011). In example (8), the Loniu sound value for 1 [si:] is combined with the English sound value of the letter p [pi:]: *sipi* 'half'.

(7) U R my 3M
'you are my drie-m'
'you are my dream'

(8) 1p
'sipi'
'half'

Consonant writing, although alphabetic, that is, segmental, shows a tendency toward logography: the sequence *txtng*, for example, no longer represents all individual sounds, but stands for an entire word. Some of these forms, such as *vs* (for French *vous* 'you') and *sry* (for English 'sorry') are widely used and pre-date the era of digital writing. Again, such strategies are not limited to English. In Taiwan writers apply the principle of consonant writing to the phonographic script Zhuyin Wen, which is used in combination with traditional logographic characters (Su 2007). In example (9), Ynnep, a Taiwanse-South African in her twenties, reduces *le* (marker of completive action) and *de* (the proposition 'of') to their consonantal value.

(9)

我	訂	蛋	糕	ㄌ	芋	頭	口	味	ㄉ	喔
wǒ	dìng	dàn	gāo	l	yù	tóu	kǒu	wèi	d	o
I	order	cake		PERF	taro		flavour		of	DM

'I've ordered the cake, it is taro flavour' (Yang 2009)

Phonetic spellings bring aspects of speech – and thus orality – into writing. They dramatize the sounds of speech, and visualize the voices associated with different accents and ways of speaking. In historical linguistics this is sometimes referred to as eye dialect. The term emphasizes the visual nature of such spellings; that is, we grasp the sound through the eye. In example (10) – from a World of Warcraft in-game chat – the French initialism *TK, tout cas*, 'anyway', is followed by a phrase representing Montreal dialect.

(10) TK MOE CHE PO
'tout cas, moi, je ne sais pas'
'anyway, I don't know' (Thorne 2012: 302)

Phonetic spellings can represent standard as well as non-standard pronunciations. Thus, the respellings *nite* and *guyz* (rather than *night* and *guys*) reflect standard pronunciation. Spellings such as *luv* and *d man* (*love* and *the man*), on the other hand, represent regionally and socially marked non-standard pronunciations (British working-class and African American English, respectively). Prosodic spellings such as *gooooood* (indicating vowel lengthening) also belong to the broad category of phonetic spellings. The example in (11) comes from Siri Lamoureaux's (2011) work on Sudanese texting (the repeated letter is <ي>, /iː/).

(11) مشتاقييييين
mushtaagiiiiin
'I miiiiiis you'

And finally there is aesthetic writing or typed calligraphy. Like futurist poets, some digital writers treat the screen as a canvas and play with the form and arrangement of letters. Example (12) comes from the Italian text message data collected by Susan Herring and Asta Zelenkauskaite (2009). The message is written in standard orthography, but the writer uses the spacebar to create a special visual effect: words are stretched and elongated, word boundaries are obscured.

(12) P u o i m e t t e r e a u d i o s l a v e g r a z i e m
'Puoi mettere Audioslave grazie mille'
'Please play Audioslave, thanks a lot'

Typographic play is a popular way of making one's messages 'strange', more decorative and artistic. For languages using the Latin script, leetspeak, or *l33t sp34k*, which emerged in the context of the gaming and hacking subculture in the 1980s, is an important resource (McKean 2002). The name itself plays on the word 'elite', referring to those who had administrator status and access privileges on early bulletin board systems. The principle that informs leetspeak is transletteration: letters are replaced with alternatives on the basis of the way they *look*. Thus, the letter 'a' or 'A' could be spelled as @, |\, or in any other way the writer thinks would be intelligible to the reader. This is different from transliteration, that is, the replacement of letters or characters on the basis of the way they *sound*. Thus, Russian Я would be *transliterated* into the Latin alphabet as <ja>, but could be *transletterated* by R or ®. The underlying principle of these letter substitutions is what the Japanologist Roy Miller (1967) has called 'total availability': anything linguistic or symbolic exists as a resource to be exploited, can be borrowed, modified and used, irrespective of its original meaning. Leetspeak-like substitution practices are no longer limited

Figure 7.6 The comic *Megatokyo*: characters 'speaking' leetspeak. (http://megatokyo.com/strip/9). Transcription: 'The pain!!!/ I need help!/ I need you to get the doctor. I got some bad pain in my chest. I need my pills!'

to specific subcultures, and Figure 7.6 illustrates the representation of leetspeak in popular culture, in the Manga comic *Megatokyo* (2000).

And as with the other strategies, leetspeak-like practices are not limited to English. Examples (13) and (14) are instant-messaging status updates by Kay, a 14-year-old Afrikaans-English teenager who lives in a socio-economically marginalized, working-class neighborhood in Cape Town (Coetzee 2012: 111). Like the 14-year-old Zambian girl mentioned by Spitulnik, Kay is fully conversant with global trends. She skillfully combines transletteration with other digital writing strategies, including consonant writing, acronyms and rebus writing (*wHr* 'where', *DC* 'disconnected', *n1* 'no one'), as well as stylized and decorative spellings (*GoEzZ* 'goes', *d@* 'the').

(13) WeN d@ w!nD bLoWzZ n1 nWz wHr K@y GoEzZ
'when the wind blows no one knows where Kay goes'

(14) M€N$€ €K M¤€+ N¤U K@VO€G@ M¥ BT !$ P@P . . . €N DC $€ M@$€
'people ek moet nou kavoega my battery is pap . . . en disconnected se ma se'
'people I must go now my battery is flat . . . and disconnected my mother's . . . '

Leetspeak-like practices are not limited to Latin script. Carmel Vaisman (2014) discusses Fakatsa, a script that combines Hebrew and ASCII letters. Thus, as illustrated in example (15), the Hebrew first person pronoun אני (*ani*, 'I') can be rendered in various ways (Hebrew is read from right to left).

(15) *Fakatsa* *Hebrew*
*נK < אני
^]א < אני
׳JX < אני
+נ% < אני

Originally, *Fakatsa* was a derogatory slang term to describe Paris-Hilton-type performances of femininity: shallow, loud, consumerist and obsessed with appearances. The term has been appropriated by young female Israeli bloggers on Israblog to describe a femininity they desire, and in online environments they display, quite self-consciously, a girlish, cute and glamorous femininity. It is not only script choice that indexes an online Fakatsa, but also the design of their blogs, which are 'dominated by the colour pink and excessively decorated with blinking or glittering kitsch, popular culture, and celebrities iconography' (Vaisman 2014: 3). Transletteration can also be found for Greek (where it is called *Greeklish* when using Latin letters to write Greek: Tseliga 2007; or *Engreek*, when using Greek letters to write English, for example, αττιτυδε 'attitude'), for Arabic (where it is called Arabizi: Palfreyman and Al Khahil 2007; Lamoureaux 2011), and on the Chinese internet (where it is known as 'Martian language': Dong et al. 2012). Another version comes from Japan, where it is also associated with girl-ness and referred to as *gyaru-moji* 'girl characters' (L. Miller 2011). Japanese writers have four scripts at their disposal: Kanji (borrowed Chinese characters, a logographic script), Hiragana and Katakana (phonographic, syllabic scripts) as well as Romanji (Latin script, phonographic and alphabetic). The replacement processes that characterize *gyaru-moji* are more complex than those for leetspeak or Fakatsa, where the visual correspondence is only between the Roman/Hebrew letter and its leet/Fakatsa equivalent. In *gyaru-moji*, replacement can be based on (a) the shape of the original sign or (b) the shape of the Roman transliteration of the character; or (c) the character can be disassembled into separate shapes that are then replaced individually (L. Miller 2011). For example, ヨ represents the syllable 'yo' in Katakana. It can be written as *E*, based on the shape of the original sign, or as *чo*, based on its Roman transliteration. In the case of composite characters both components can be replaced individually on the basis of their visual resemblance to other symbols. For example, the composite symbol は ('ha', Hiragana) can be pulled apart and rendered as レよ, using the Katakana symbol for the syllable 're' and the Hiragana symbol for the syllable 'yo'.[6] The degree to which a writer employs *gyaru-moji* is associated not only with the projection of a girl-identity, but also with emotional intensity. Sending a message that contains an abundance of graphic substitutions is considered to be a sign of affection and intimacy because of the time it takes to input the decorative spellings.

Brevity has been described as a central feature and maxim of digital writing (Thurlow 2003). Brevity is a design feature of certain interfaces (SMS and Twitter, which limit the number of characters), facilitates quick replies in interactive genres such as chatting, and echoes the Futurist 'aesthetics of speed'; an aesthetic shaped – as noted in the previous section – by the telegram and the postcard, rather than the letter and the book. Yet not all of the strategies discussed here reduce letters and allow writers to produce texts quickly. Aesthetic writing in particular can take time and often produce words that are as long as, or even longer than, the standard orthographic form. Instead of brevity and communicative economy, they emphasize the visual, multimodal aspects of writing and the writer's creativity (much like the heteroglossic practices discussed in Chapter 6). Especially in low-income contexts where access to computers remains limited and mobile phones are often fairly basic, creating multimodal texts means typed calligraphy – the artful manipulation of the visual side of language – rather than mash-up videos or remixed images.

MULTILINGUAL WRITING: VOICING CONTRASTS

The previous chapter looked at how writers use their heteroglossic resources to project, perform and style different personae online, and to articulate stances toward what is happening in the interaction. Among the resources people have are not only different languages, varieties and styles, but also different ways of representing language visually. Writers thus make several decisions when composing a text. First they need to decide which linguistic forms to use, and then how to spell them. Example (16) presents a short exchange between Sibo and Sxosh, a South African couple in their early twenties. Both speak English as well as isiXhosa (isiXhosa in italics).

(16) Sibo: Sori babes *kuphele umoya* talk 2 u 2moro
'Sorry baby airtime got finished talk to you tomorrow'

Sxosh: Ok *ndisenawo* bt talk 2 u b4 *ndilale*
'OK. I still have some but talk to you before I sleep.' (2011)

Sibo and Sxosh use global conventions when writing English. There is rebus and consonant writing as well as a phonetic respelling of 'sorry' (indicating vowel lengthening before /r/). The only English word spelled according to standard orthography is 'talk'. IsiXhosa, on the other hand, is consistently spelt in full. What we see in this short and ordinary exchange are not only two languages, but also two different ways of representing these languages visually. Sibo and Sxosh combine English and isiXhosa syntactically, but keep them apart orthographically.

Volker Hinnenkamp's (2008) work on German-Turkish digital writing by diasporic youth describes a similar situation. In his data, German is not only the language that is quantitatively dominant, but also the language that is orthographically manipulated. The bilingual Turkish–German writers use various phonetic spellings to indicate local varieties of German, and creatively draw on Turkish spelling conventions to fashion a new-look German that stylizes and visualizes migrant accents. Turkish, on the other hand, although used in a colloquial style, is written in conventional orthography. Why are the two languages treated differently? According to Hinnenkamp, it is precisely the dominance of German that makes it a target of appropriation through creative manipulation: 'taking possession of it in one's own sense and nobody else's' (p. 271). In other words, the very dominance of the language encourages its subversion, whereas the socio-politically fragile situation of the other language demands linguistic, in this case orthographic, respect.

The idea that *respect* can be accorded to a language and that 'correct' spellings are a visible sign of such respect is supported by our South African data, and to abbreviate African languages was felt by many to be 'very wrong'. Linda, a 14-year-old isiXhosa-English student, comments as follows:

(17) I abbreviate English because I have no respect for the language, however, I respect isiXhosa and isiZulu and so do not abbreviate them. (Questionnaire response, Cape Town, 2011)

Not everyone is as metapragmatically explicit as Linda, but the overall pattern is fairly stable. Although there exist some abbreviations – such as *GPY*, mentioned above as a play on *ROFL*, *CC* for *sisi* 'sister', and *mf2* for *mfethu* 'friend' – isiXhosa, like Turkish, is usually written conventionally. The only common modifications are spellings that reflect connected speech phenomena and colloquialisms. But visually, isiXhosa digital writing does not usually disrupt the reader. The situation is similar for Wolof in Senegal (Lexander 2011) and Giriama in Kenya (McIntosh 2010). However, this observation cannot be generalized for African languages or minority languages in general. Kiswahili, politically fairly powerful and used as a lingua franca across East Africa, is commonly abbreviated (Barasa 2010; example (6)), and the same is true for Loniu, discussed above (example (8)).

Whether one abbreviates freely or spells out in full can have consequences for the actual writing process. While English messages are composed 'on the fly', with considerable tolerance for orthographic variation, misspellings and abbreviations, texting in, for example, isiXhosa requires careful attention and editing. Given that many writers do not study African languages at school, but are taught through the medium of the former colonial language, writing African languages 'correctly' is a challenge as well as a source of pride. One person who took part in the chat experiment described in the previous chapter, umgqusho, compares her digital writing in English and isiXhosa:

(18) It took me three times as long to type the Xhosa version, 'cause, 'cause, with English it's natural, whereas with vernac it's like, is there an 'n'? Even when you are checking, you are unsure, you have to concentrate, with English, whatever, they'll know. (Interview data, 2008; Deumert and Lexander 2013)

The expression 'whatever' in example (18) is reminiscent of Naomi Baron's (2008) argument that online writing is characterized by 'linguistic whateverism'. Whereas writing in isiXhosa requires close attention to the norms of standard orthography, writing in English allows for a great deal of freedom and *laissez-faire*. The comment by umgqusho shows that such 'whateverism' can be applied selectively to certain languages, but not to others. This selective application creates what Asif Agha (2005) has called voicing contrasts. The two voices are clearly distinguished from one another, both formally and indexically. The English voice is modern, cool, and ultimately not concerned with 'getting it right'. It no longer evokes the old colonial standard, the language of the classroom, but rather a global, transgressive postmodernity. The isiXhosa voice is serious and respectful, is deeply local and displays rootedness in tradition.

Such voicing contrasts can be used strategically. Example (19) is a late-night text message sent by Nandi, in her early twenties, to her boyfriend. As in example (16), the message combines not only two languages (English and isiXhosa), but also two different styles. *Luv ya* is an example of a phonetic spelling, projecting a stylized, informal accent; *Radebe wam*, literally 'my Radebe', uses the boyfriend's clan name, and constitutes a traditional and respectful form of address (similar to the salutations discussed in Chapter 6; see also Deumert and Masinyana 2008).

(19) Luv ya Radebe wam (2008, South Africa)

The juxtaposition of *luv ya* and *Radebe wam* is more than a juxtaposition of languages and writing conventions; it also juxtaposes two ways of expressing affection, that is, playful, informal and flirtatious versus serious, respectful and formal. Similar patterns have been described for Francophone Africa. In Senegal, French – abbreviated and modified – is commonly found in the playful role, while mature and serious love is carefully composed in Wolof (Lexander 2011). Voicing contrasts have also been reported for Hong Kong, where writers move into classical Chinese to enhance the emotionality of a text (Lin and Tong 2007). 'I love you' messages are a difficult genre to master. Expressing one's love means – again and again – overcoming the 'fatigue of language' (Barthes [1977] 2002: 20), that is, putting into words the specialty and uniqueness of one's desires. Judith Butler (2011) describes *I love you* as a deeply problematic speech act and asks: how can we convince someone that we really do love them, that we are not just repeating a stock phrase, a cliché, 'that belongs to no one and to anyone'? By juxtaposing linguistic forms and their indexicalities, writers are able to move beyond the 'anonymous citationality' and blandness of *I love you*. The message becomes multi-voiced and as such unique.[7]

But what if you are not multilingual and write in 'one language' only? How do you express the difference between being serious and being playful? Between flirting and loving? Orthographic representation matters here too: if you really mean it, make the *effort* to spell it out. Example (20) comes from a discussion between Sharon, Etienne and Ralph about how to say 'I love you' in a text message or when chatting online (2010). All three are English-dominant South Africans in their late teens.

(20) Ralph: I know there is the L-U and then you get like L-V-U, LV space U
Etienne: it's a, like different degrees of loving
Sharon: the whole word is actually like love, that is, like love when you mean it, you write the full *I l-o-v-e y-o-u*
Ralph: when you speak more proper, more formal language
Etienne: you mean it more, depends on who you say it [to], you also make jokes like 'I love you man'
Sharon: but then you don't write the whole word

The maxim 'if you mean it make an effort' is not limited to South Africa. Luanlegacy (from the US) complains in his video 'Texting Etiquette' (2011) about people who send very short messages, such as *k* ('okay').[8] In her comment in example (21), pieakatt, also from the US, discusses this with reference to 'I love you' messages, a speech act where abbreviations are just not cool.

(21) Your talking to someone you love whether it be your boyfriend or family or friends or whatever.and you take the time to say I love you . . . and they send back . . . "good cuz i luv u 2" bitch you can't spell out because, capitalize 'I', and spell out love, you, and too . . . fuck you! (posted in 2011)

The opposite maxim also holds. Sending extremely short messages can be a sign that one is unhappy and upset. Sharon explains this as follows: 'the shorter the words, the sentence that you make, they know that there is actually something wrong' (focus group data, 2010). This brings us back to the notion of brevity. While brevity and speed are desired at times, slowing it down and spelling it out can be just as important. And it is not necessarily about spelling correctly. Non-standard messages can be 'special' too if they are composed carefully and creatively. The text message in example (22) was sent by Trevor, an Afrikaans-dominant male South African, to his cousin, Chantelle, who is female and English dominant. She appreciated the message not only for its sentiment and the creative expression of affection (*mwa-hugzikissiba*), but also because he made an effort to write to her in her preferred language (even though he struggles with it).

(22) Mwa tweetie, happy v.nite . . . h0pe ur n0t inlove 2day, but be inl0ve everyday! Mwahugzikissiba (^^,), HAVE A NICE V.NIGHT SWEETY. Lovies (2011)

The message, carefully composed, can thus become a gift, and similar messages are sent back in acknowledgment and appreciation. The various transletteration practices discussed in the previous section, namely leet speak, Fakatsa and *gyaru-moji*, are examples of such careful and creative composition. And although originality is valued, prefabricated messages – which can be downloaded from the internet, copied from books or purchased from service providers – are also treasured, just as we might treasure a mass-produced greeting card. And in the same way as we personalize the mass-produced greeting card with our own, handwritten message, so do writers modify and personalize prefabricated messages. They rarely forward the message as it is but will typically personalize the spellings, even if just minimally, and add forms of endearment (on texting as 'gifting' see Taylor and Harper 2002 for the UK; Ito and Okabe 2005 for Japan; Lin and Tong 2007 for Hong Kong; Ling and Yttri 2002 for Norway; Lamoureaux 2011 for Sudan; Archambault 2012 for Mozambique; on prefabricated messages as 'gifts' see Ellwod-Clayton 2005).

BETWEEN CONVENTION AND CREATIVITY

Non-standard digital writing exists as a fairly well-delineated object in the popular imagination. It is referred to by names such as *textspeak, weblish, chatlingo, netspeak* or *textese*, and codified in dictionaries and online translators. Consumer culture provides us with T-shirts, mugs and baseball caps that display examples of leetspeak or acronyms such as *OMG*, and advertisers draw on such forms in promoting their products (as, for example, in the case of Calvin Klein's 2007 *in2you* campaign). Even those who oppose these practices are able to emulate them in order to mock them: 'I'm always glad to say that I text with words spelt full, instead ov tlkn lyl ths' (YouTube comment, 2013, in response to the video 'Texting is Ruining our Language!!').[9] Such representations, and the metapragmatic discussions that accompany them, contribute to what Asif Agha (2003) has called enregisterment. That is, digital writing is perceived as a distinctive and conventionalized register or style, which is expected, or at least licensed, in specific contexts (Squires 2010).

But how common are these strategies actually? Do writers use them a lot? Or do they just draw on them here and there? With respect to English, linguists have

Table 7.2 Variation across corpora and countries: *u* instead of *you*.

Country	Source	Percentage of *u*	Data
United States	Squires (2010)	7.4	IM
Canada	Tagliamonte and Denis (2008)	8.6	IM
United Kingdom	Tagg (2009)	39.0	SMS
South Africa	Deumert and Lexander (2013)	68.0	SMS
Ghana	Deumert and Lexander (2013)	75.0	SMS
Nigeria	Deumert and Lexander (2013)	88.0	SMS

Note: Comparing texting data with IM data is not ideal given the different affordances of phone keypads and computer keyboards. However, Ling and Baron's (2007) work on US texting suggests that although abbreviations and shortenings are more common in texting than in IM, frequencies remain low (below 4 percent in their text message corpus).

argued that such forms occur at fairly low frequencies in actual data sets (see, for example, Ling and Baron 2007; Crystal 2011: 4ff; Tagliamonte and Denis 2008; Squires 2010; Thurlow and Poff 2013). Table 7.2 summarizes the results for one common and long-standing variable: *u/you*. Currently available quantitative data suggests that this form is rare in North American data, more common in the UK and ubiquitous in Anglophone Africa. Similar patterns apply to a range of other variables, and Anglophone African texting generally shows a high density of non-standard spellings (for statistical details see Deumert and Mesthrie 2012; Deumert and Lexander 2013). In example (23) an American exchange student at the University of Cape Town comments on what she perceives as a deeply local practice.

(23) My first text message from a boy from Zimbabwe read: hw ws the rest of ur nyt? Only gt hme @ 6 – im hrtng. Wnt 2 grab breakfast wit me 2mrw mrning? Wil cal u l8r wit details. I have never seen this style of texting from any American and after asking my friends, apparently this is how they text message here. (2010)

The orthographic principles and strategies of the genre – rebus writing, substitution cyphers, abbreviations, phonetic spellings and aesthetic writing – are thus realized differently in different places. They are not simply a replication of a global (English) norm, but become local practice (Pennycook 2007; Blommaert 2010). This applies not only to frequencies, but also to the indexicalities of these forms, that is, form–meaning relationships. Consider transletteration practices: these can index – depending on context and locale – geekiness, insider status, girlishness, modernity, but also a sense of being retro and even uncool: 'My brother thinks it's cool to text me in leetspeak (no)' (Twitter 2013). In addition, some forms might be well known in one place, but puzzling in another. Examples of this are the forms *mic* and *xo* in South Africa. In South African text messages *mic* does not usually refer to a microphone, nor does *xo* represent a kiss and a hug. Instead they are 'stylish' respellings of the verb 'miss' and affirmative *sure* (pronounce locally as [ʃo:]; Deumert and Masinyana 2008).[10]

An essential aspect of the poetic is its fundamentally open-ended nature in both form and meaning. Yet it is precisely this openness that has often been downplayed by linguists, who – in the time-honored tradition of the discipline – attempt to show digital writing, like any language or variety, as structured and systematic, not wild, anarchic and unpredictable. Caroline Tagg (2012: 51), for example, draws on Mark Sebba's (2007) discussion of non-standard spellings, where he argues that 'even unlicensed variation from the norm is constrained to a large extent by them'. Tagg then considers two different spellings of 'school', *skool* and *zguul*, and argues that the second spelling is

> not a meaningful deviation from the spelling of *school* because it does not follow English orthographic principles: <zg> is not a permitted sequence . . . while <uu> is very rare . . . *<Zguul> has no meaning at all.* This shows that respellings are neither freely or randomly chosen, but that choice is restricted according to orthographic principles. (My emphasis)

Similarly, Jan Blommaert (2012b: 8) characterizes digital writing as strongly normative: 'it is a system, something that operates on the basis of quite rigorously applied rules, deviation of which is possible but never unlimited and always comes at a price'. Such statements about the ordered and systematic nature of digital writing make it difficult to explain forms such as *mic* or *xo*, since English orthographic principles don't allow us to represent [s] as <c>, or [ʃ] with <x>.

But do respellings actually have to be principled? Can we not imagine a situation, a specific context, where *zguul* could become quite meaningful, even though – or maybe precisely *because* – it defies the principles of English orthography? Writing online, as noted above, is not about *rules*, but about strategies and the display of creativity. Following the postcolonial theorist Achille Mbembe (2004), we can describe these practices as going beyond mere mimicry, simply imitating and copying pre-existing forms; rather they are about mimesis. That means that similarities are established with existing forms, and existing patterns are exploited, but at the same time writers are able to break through existing molds and invent forms that are unpredictable and original.

The openness of the system is clearly visible in the South African questionnaire data, where more than two-thirds of respondents said that they create 'their own' initialisms. Examples include: *IHY* ('I hate you'), *e^n^u* (*en nou nou* 'and now', Afrikaans), *wcwca* ('what can we chat about'), *HAAEMB* ('help ants are eating my babies') and *cqtm* ('chuckle quietly to myself') as well as a range of forms they were not willing to share ('lyk rude stuff').[11] Such creations can be difficult for readers to decipher. Initialisms introduce ambiguity into the interaction – a hallmark of poetic language – and upset the Gricean maxim of manner (express yourself clearly, avoid ambiguity and obscurity; Grice 1975). And indeed, sometimes readers struggle to understand the texts they receive. Close to 80 percent of South African questionnaire respondents indicated that they had experienced situations where they could not decipher a message. Nenagh Kemp (2010) reports similar results for Australia and finds – in line with several other studies – that while many of the strategies discussed in the previous section accelerate the speed of writing, they inevitably slow down the reading process. As one of the British teenagers put it in an early study by Rebecca Grinter and Margery Eldridge (2001): 'you have to sit there thinking l-8-r, or oh, later . . . '. While *l8r* is now well established and processed quite automatically, other forms might still leave us with challenges.

How do you learn a way of writing that is not rule-governed but open? How does the novice become an insider and skillful user? Learning takes place not through adult authority figures such as parents and teachers – who are the gate-keepers of standard literacy – but rather through peer networks, and also, importantly, media practices. In some countries, such as South Africa and Mauritius, TV programs, which screen text or Twitter messages from viewers, have been an important resource for learning about alternative spellings. For those who have access to the internet there are dictionaries and online translators. Printed booklets such as *Climax of Love and Romantic Text Messages* (Nigeria) and *Text Me 2Night* (Ghana) are also popular (personal communication, John Singler). And, of course, every message received, every chat conversation is a chance to see and to learn new ways of manipulating language. *Ufunda ngobona*, isiXhosa for 'you learn by observing', is a common answer in our South African interview data. However, there are also *moments* of instruction, and sometimes we might ask those who we consider more knowledgeable for help. This is particularly common in parent–teenager interactions, where teenagers are cast in the role of the 'digital native' and thus are the teachers and gate-keepers of digital literacies (while adults remain gate-keepers of conventional literacies). The text message exchange in example (24) between a mother and her teenage daughter was posted as a screenshot on Twitter (by the mother; United States 2013).

(24) Daughter: Mmk! Sounds good.
Mother: I'm not sure what mmk means :-)
Daugher: It means ok. I say that in text a lot :-)
Mother: Oh, okie doke!
Daughter: Haha
Mother: Or shall I say okd :-)
Daughter: You shall say okie doke. ;-)

The mother commented on this exchange in her Twitter status as follows: 'So much for trying to be cool and using textese with my teenager'. Yet the mother's proposed form 'okd' is not 'wrong' or 'impossible'. It uses common strategies, but just does not get the approval of her daughter. And indeed, sometimes teenagers feel quite embarrassed by their parents' enthusiastic digital engagement. While some complain that their parents simply don't get it and write in a style that is too formal, others are confused by their parents' tendency to 'over-abbreviate', and portray them as akin to children who are as yet unable to control the new toy (South Africa, focus group data, 2011). Nomhle, 22 years old, describes – with astonishment and some exasperation – her mother's unexpected creativity: 'my mom would say, for example, she wants to say "I crossed the road", and she would use the cross symbol with a *d*, and you would have to work it out' (2011). *I xd da rd* – why not?

CONCLUSION: LIQUID LANGUAGE

Drawing on Zygmunt Bauman's notion of liquid modernity, Oren Soffer (2012) describes digital writing as liquid language, immensely pliable, flexible and fluid. However, while individual creativity is licensed, the similarities, within and across languages, suggest that there exist broad community *preferences* as to what is considered 'skillful' digital writing. Nandi, a 19-year-old South African from Mthatha, does not hesitate when asked how she would define a 'good texter'. To her it means writing quickly, playing with words and knowing one's abbreviations (Deumert and Lexander 2013). Stanley Lieberson (2000) has argued that tastes and fashion – unlike art, which is original and singular – are always shaped by collective processes. That is, in the here and now, individuals are making choices, which are embedded in broader social and cultural, rather than purely individual, preferences and tastes. Ornamental, non-standard writing, which emphasizes the visuality of language, has a broad appeal, a long history and multiple points of origin. Artistic movements such as Futurism helped to establish an aesthetic that was picked up and democratized by popular culture from the 1960s onwards. This interacted with various vernacular practices such as children's early handwriting, substitution cyphers and secret languages, decorated hand-written letters and notes that are passed around in class.

Digital writing follows conventions but also allows us to transcend precisely these conventions. We can develop new forms by playing around with existing strategies such as letter substitutions, rebus writing or phonetic spellings – all of which are examples of poetic language in the sense of Jakobson. Unlike writing in the standard, digital writing is both conventional and playful, exploiting the visual side of language to maximum effect. Innovative forms that 'make strange' are always possible. Speakers experience this as freedom from the strictures and surveillance of the classroom. Mpale states this emphatically in Sesotho: *matsatsing ana re ngola joalo ka ha re batla mos, pene ea tichere e khubelu mona e absent, ke eao o joetsa!*, 'these days we write as we like, the teacher's red pen is absent here, I tell you!' (Lekhanya 2013).

The texts that such strategies produce are not uncontested. Not only do parents and teachers, the gate-keepers of principled, ordered and normalized literacies, express concern about non-standard spelling, but there are also those among the young and technologically savvy who reject such spellings and express counter-tastes. In these discourses, we see a strong desire for norms and correctness, for a writing that does not disrupt, that is, a zero-tolerance approach to non-standard forms (25).

(25) I HATE texting. Not only so many do it (Trust me I have seen people in IRL who do this) it makes me want to kill this generation of teens because of this. In the future i'm pretty sure only "Normal" people will write

and type correctly. NO MOAR BFF,OMG AND DA LAWLS PLZ.
(Comment to the video 'Texting is Ruining our Language!!', 2013)

NOTES

1. Digital writing also shows grammatical modification, and syntactic ellipsis is particularly common. Useful discussions are found in Henn-Memmesheimer and Eggers (2010), Herring (2012) and Tagg (2012). While grammatical modification is not a particularly strong feature of interactive digital writing, it can be prominent in some sub-genres. An example are lolcats, i.e. photographs of funny, cute or grumpy cats with a superimposed humorous caption. The language in which the captions are written – called lolspeak or 'kitty pidgin' – combines grammatical features and stylistic forms closely linked to baby-talk (over-generalization of regular inflections, lack of interrogative inversion, reduplication, unanalyzed phonetic strings such as *aifinkso* 'I think so', and transparent word-formation; for discussions see Gawne and Vaughan 2011).
2. Although Jakobson's work has been positioned in opposition to Bakhtin (e.g. Sturrock 1986), I follow Kristeva ([1974] 1980) in seeing their work as complementary.
3. A full bilingual edition of *Calligrammes* is freely accessible at http://books.google.co.za/books?id=7FviP1Cl3jcC&printsec=frontcover#v=onepage&q&f=false. The translation is by Anne Hyde Greet.
4. Alternative categorizations are offered by Androutsopoulos (2000), Sebba (2007) and Tagg (2012).
5. For English, Partridge's *Dictionary of Abbreviations* (1942) lists pre-digital examples, many of which continue to be in active use, e.g. *FYI, RIP* (Crystal 2008: 47). Others, however, have fallen out of use (such as *SWALK*, 'sealed with a loving kiss', which was common at the end of love letters). An even earlier example is the 'Boston abbreviation craze' of 1838 described by Read (1963).
6. The use of Latinized scripts can also be a question of affordances rather than choice. Writing from a computer usually provides one with the choice of either script; writing from a phone often restricts writers to Latin letters only (Tagg and Seargeant 2012). Moreover, in the diaspora, many writers have not acquired the non-Roman script and habitually represent their spoken vernacular through the locally used system, usually the Latin alphabet (Androutsopoulos 2006).
7. Voicing contrasts and parallelisms of this type are a general feature of artful language. Consider Robert Burns' poem 'A Red, Red Rose' (1794)' which combines literary-style standard English with Scottish dialect.
8. Until mid-2013 the video was available at http://www.youtube.com/watch?v=gmjiPSo2-BQ. The account has since been terminated 'due to multiple third-party notifications of copy-right infringement'.
9. http://www.youtube.com/warch?v=c0UVU585VZM.
10. The use of *x* to present [s] or [ʃ] – rather than its sound value [eks] – is also found in German (*sonx*, 'songs', [sɔːŋs]; Androutsopoulos 2000) as well as in Portuguese texting (*xim* for *sim*, 'yes'; Silva 2011). In our South African data we also see forms such as *xul* for *school.*
11. The questionnaire data (N = 554) was collected in 2010 and 2011. The majority of respondents are first year university students (aged between 18 and 21 years) and high school students (aged between 13 and 17 years).

Chapter 8

Sociability online: between *plaisir* and *jouissance*

> No sooner had I arrived here than I knew I found a place called home. I want to virtually grow old with some of you people. You're real.
>
> Twitter 2013

INTRODUCTION: HANGING OUT REVISITED

Chapter 3 discussed the work by Ito and her colleagues (2010), who identified the leisure activity of hanging out as central to the digital engagement of young Americans. Digital hanging-out practices were also shown to be common among those with less connectivity. Facebook (founded in 2004), especially, has captured the global imagination and, in 2013, was the top social network(ing) site in Africa, Asia, South America, Europe and North America. It has overtaken Mixi in Japan, and the once-popular applications Orkut and Badoo in South America. The only areas where it does not dominate are China and Iran (where online censorship is enforced), and the former Soviet countries where Odnoklassniki (Одноклассники) and VKontakte (ВКонтакте) are alternative network(ing) sites.[1]

Hanging out can be intimate and private, restricted to insiders and unwelcoming to others. However, hanging out can also be public or semi-public, and the focus in this chapter is on the latter type. The sociologist Ray Oldenburg (1989) refers to the places where sociable interactions occur as third places, and he argues that informal public gathering places are important for a healthy social life. Third places are not located in the intimacy and privacy of home, nor are they connected to the seriousness of work, where interaction is typically utilitarian and purposeful. Oldenburg's work draws on the sociology of Georg Simmel and, especially, his writings on sociability, which emphasize the conversational enjoyment and sheer pleasure in the company of others that characterize social interactions in such contexts. Thus, in third places, the overall mood is convivial and interactions are experienced as pleasant and fun, creating a sense of togetherness and good fellowship. Examples of third places include taverns, coffee-shops, street corners, markets and village stores as well as – in the virtual realm – chat applications such as IRC (internet relay chat), Facebook, Twitter, Second Life and so forth (for an early discussion of the virtual as a third place, see Rheingold 1993).

This chapter takes a closer look at the enjoyment we experience when we engage in sociable interactions. In doing so I draw on Roland Barthes' *The Pleasure of the Text* (1975), where he follows the psychoanalyst Jacques Lacan in distinguishing two main types of enjoyment: *plaisir* and *jouissance*. Socially embedded enjoyment, as described above, usually gives rise to *plaisir*, an uplifting, compliant, comforting and cheering 'pleasure' that strengthens social bonds and creates community. *Plaisir* 'comes from culture and does not break with it'; that is, it stays within the norms of acceptability and does not transgress. But not all social interactions, offline and online, are comforting *plaisir*. In 2013, for example, news sites and blogs were discussing 'Facebook's big misogyny problem', that is, postings of distasteful, sexist and crude content.[1] Such debates draw attention to the dark underbelly of the internet. Online spaces not only allow for wholesome, pleasurable interactions, but are also unmonitored expressive spaces that can provide a forum for sexists, racists and other anti-social groups. While misogynistic and racist texts can be posted aggressively to offend or as part of an overt political agenda, their presence in online conversations can also be reminiscent of the way crude, vulgar and offensive jokes are told and laughed at in schoolyards, daring and eliciting a mix of nervous laughter and shock among those who are part of the in-crowd. This is, however, an unsettling kind of enjoyment, known to be wrong, inappropriate and offensive, but there is laughter nevertheless. The tone is different from that of *plaisir* and moves closer to Barthes' second form of enjoyment, namely *jouissance* (which does not have a straightforward translation equivalent in English). *Jouissance* stands in direct opposition to Sigmund Freud's 'pleasure principle', that is, our desire to seek pleasure and to avoid pain. *Jouissance* refers to a type of enjoyment that lies at the border of pleasure and displeasure. Thus, unlike *plaisir*, *jouissance* is transgressive enjoyment, often sexualized, always subversive, discomforting and ultimately disruptive to the social order (see also Deumert forthcoming b). It is similar to what Bakhtin ([1965] 1984) calls the *carnivalesque*: moments in social life where conventions and the traditional order are set aside, and the routines of daily life are, temporarily, suspended.

THE PLEASURES OF *PLAISIR*: COMMUNITIES AND SOCIABILITY

Community, its conceptual fuzziness and lack of theoretical precision notwithstanding, remains an important emic term that people use to describe an experience of fellowship and togetherness. Raymond Williams ([1976] 1983) describes it as a 'warmly persuasive word' that 'never seems to be used unfavourably' (p. 76). Discussions of community often carry with them folk notions about ideal forms of social life. Villages and integrated neighborhoods where people know one another and interact frequently are prototypical examples of community. A number of sociologists – such as Robert Putnam (2000) in *Bowling Alone* – have argued that communities of this type have disappeared and that personalized forms of media consumption have led to growing social isolation and disconnection. We used to go bowling together and borrow sugar from our neighbor, now we just sit alone in front of television and computer screens. Social and socio-psychological research, however, has shown that a sense of community and a feeling of belonging can develop online, and debates as to whether 'virtual communities' are 'genuine communities' are beginning to have a dated feel to them. Modality and physical presence do not seem to affect whether we develop feelings of trust and mutual obligation, security and belonging (Bargh and McKenna 2004; see also Chapter 2 on the reality of the virtual; Chapter 3 on communities of practice and affinity spaces; Chapter 5 on discourse communities). In other words, far from reducing social connectivity, mobile media afford us new ways of being together. We might not be bowling together, but we are chatting together; we share photos, provide advice and support, laugh and flirt, break up and make up.

Online and offline interaction do not exist in isolation from one another, and people use whatever modalities are available to communicate: body-to-body when possible; telephone, letters and digital media when apart. In these contexts, a global platform such as Facebook can become a place for locally embedded practices and meanings. This is illustrated in example (1).[2] The topic of the brief Facebook (2012) interaction is Moegamad's lunch of *daaltjies*, an Afrikaans expression for what are called *chilli bites* elsewhere. All participants reside in Cape Town and are Muslims in their thirties and early forties. The exchange took place just before the start of Ramadan.

(1)

Status update		
Moegamad	My lunch from the food market.	Accompanied by a photo of two chilli bites on a plate.
Comments		
Aeysha	that's a little	
Moegamad	It filling alhamdumillah	'praise to God'
Moegamad	And very yumy	
Lameez	iet al klaar agte die bak	'already eating behind the bowl'
Moegamad	Ai ek sal nooit wen met jou nie how's haadiyah?	'Ah, I can never win with you'
Gadjia	miriams daljies?	
Moegamad	Nope . . . They have a open food market at st georges mall every Thursday from 11-3 pm	

The interaction is permeated with local references. Friends are mentioned (*haadiyah*), and physical places such *st georges mall* in downtown Cape Town anchor the conversation, its virtuality notwithstanding, in the concrete, the actual (Chapter 2). The language too is decidedly local, a vernacular that moves seamlessly between English and Afrikaans. The expression *iet al klaar agte die bak* translates literally as 'already eating behind the bowl' (with *eet* [eət], 'to eat' spelt according to local pronunciation as *iet* [i:t], and *agter* without the final /r/, reflecting casual speech). It refers to someone who is not taking the fast seriously and is known to eat surreptitiously, that is, 'behind the bowl'. It is a form of teasing, and understanding the phrase requires knowledge not only of Afrikaans, but also of Muslim religious practices and the kinds of jokes that can be made. Following Quentin Williams and Christopher Stroud (2010: 40), we can describe the extract as an example of 'extreme locality'. The participants are united in a deeply local version of life and evoke the 'bric a brac of events and reference points they share, and the people they know'. Moreover, the humorous key in which the interaction is conducted creates the kind of enjoyment Barthes calls *plaisir*. There is no transgression here, just the jocular affirmation of existing social bonds.

Interacting with others through digital media is experienced as deeply enjoyable by many writers.[4] Taryn, a 15-year-old Afrikaans-English-speaking teenager from Cape Town, told me how she spends hours every day chatting to others on her phone, starting as soon as she gets up:

(2) [I start as soon as] I wake up, just to get like a light in my eye, so that I can actually wake up, maybe there is one or two people on to tell them, so now I am getting my hands moving so that I can actually feel awake, it's just five minutes and I go off again, then after school maybe an hour, and a few hours before I sleep. (2010)

It is not just teenagers who spend a lot of time socializing online. Twenty-three-year-old Shaida also chats throughout the night. Laughingly, she admits that even when at work, she regularly disappears into the bathroom to continue her conversations: *incokw'imnandi, kumnandi nj'ukuncokola*, 'chatting is nice, it is nice to chat' (isiXhosa interview data, 2011).

In considering sociability – whether offline or online – the work of the sociologist and cultural philosopher Georg Simmel (1858–1918) is important. Simmel wrote widely on the sociology of leisure, about travel, food, fashion, urban life, social etiquette, money and even flirting. A text that is relevant in the context of this chapter is an essay titled 'The Sociology of Sociability' (originally published in 1911; quotes below are based on the 1949 translation). Simmel defines sociability as a special type of public and semi-public association and togetherness, where interactions are playful and pleasurable rather than useful and serious.

> The decisive point is expressed in the quite banal experience that in the serious affairs men talk for the sake of the content which they wish to impart or about which they want to come to an understanding – *in sociability talking is an end in itself* (p. 259; my emphasis).

Oldenburg's work on third spaces draws directly on Simmel's discussion of sociability as convivial social interaction that is located at the boundary of public and private. A complex mix of public and private is also a design feature of many social network(ing) sites. Although private messages can be sent, a prominent space on any social network(ing) site is the semi-public one where our posts can be seen by all those connected to us as 'friends' (Facebook), 'followers' (Twitter) or 'contacts' (mobile IM).

A core feature of sociability is the enjoyment of the company of others and a broadly playful, conversational mood. It is interaction for its own sake – talking for the sake of talking – rather than to achieve a particular goal. Sociability comes with its very own interactional maxims and etiquette: talk should be interesting and entertaining; participants need to be willing to change topics quickly and smoothly in order to keep the conversation flowing; for one person to dominate the floor is not acceptable – there should be give and take, a multiplicity of voices (heteroglossia as discussed in Chapter 6); everyone has equal access to the floor; talk should be 'light', never too heavy or too serious, too personal or too intimate. What is required is a commitment to conversation and engagement: not to respond, to just drop off, is not acceptable.[5] The aim is to keep the conversation flowing. Nosipho, a 16-year-old South African teenager, speaks appreciatively of the type of interaction she considers typical of a 'nice' chat. She explains, in isiXhosa, her own experience of conversational flow as follows: *hayi kaloku iyaqhubekeka kalok'incoko iqhubekeke, iqhubekeke, iqhubekeke, iqhubekeke* ('no man, it's like, the conversation continues, it continues and continues, and continues').

Simmel describes sociability as a special 'sociological formation corresponding to art and play'; a form of interaction that is shaped by an 'artistic impulse'. This formulation links back to the discussion of verbal art and the poetic in Chapters 6 and 7. Following Richard Bauman (1972: 341), we can position sociable events as 'belonging analytically to the realm of *social art*' (my emphasis). If we see sociability as a pleasurable but also aesthetically marked form of social interaction, then there will be those who excel at it, that is, 'good talkers' who are valued and appreciated for their conversational abilities. This is also true for interactive digital writing. Kristin Vold Lexander and myself (Deumert and Lexander 2013) have suggested the expression 'textual linguistic dexterity' to describe digital writing skills that are appreciated by conversational partners: fast, dexterous finger movements allow for a quick back-and-forth, for conversational flow, and linguistic virtuosity – the ability to use language artfully and creatively – produces texts and turns that are enjoyable to others

Examples (3) and (4) illustrate the playful sociability typical of many digital interactions. In example (3) Bad_Ass_Sinorita's arrival in the South African IRC channel #hangout initiates a range of flirtatious and humorous actions and exchanges.

(3) IRC chatroom #hangout at ZAnet (2008).

Bad_Ass_Sinorita has joined #hang-out
@ChOcO_RoMeO sits on Bad_Ass_Sinorita's lap
<@Bad_Ass_Sinorita> he baby:)
<@ShaGuaR> Lol
<@Sweet_MoE> bassie!!!!!!!!!!!
@ChOcO_RoMeO snuggles with Bad_Ass_Sinorita
<@Bad_Ass_Sinorita> Sweet_MoE!
@ChOcO_RoMeO purrssss
@Sweet_MoE huggles Bad_Ass_Sinorita
<ShaGuaR> Sup Bad_Ass_Sinorita
<@ChOcO_RoMeO> ShaGuaR dont talk to my chick
<@ShaGuaR> Arent you supposed to be going to hillfox
<@ChOcO_RoMeO> i am just now
<@ChOcO_RoMeO> have to wait for an order quick
<@ShaGuaR> Lol ok.
<@ChOcO_RoMeO> werent u suppose to meet me at newscafe on friday night at 8pm?
<@ShaGuaR> Sup Bad_Ass_Sinorita
<@ShaGuaR> Lol
<@ShaGuaR> Werent you supposed to be born a female?
<@ChOcO_RoMeO> were you suppose to be born with arches on ur feet?
<@ShaGuaR> ROFL
<@ShaGuaR> BWAHAHAHAHAHAHAHAHA

The conversation in example (3) is a typical example of sociability as described by Simmel: playful, teasing and punctuated by laughter. There is mock verbal combat and flirtation. Whether or not ShaGuaR and ChOcO_RoMeO actually had an arrangement to meet at News Café (a South African coffee-shop chain), or whether this is part of the 'weren't you supposed to' banter, is irrelevant as long as the interaction is enjoyable and provides amusement to participants and audience.

Example (4) is similar in tone. It comes from a Facebook wall and starts with Thando recalling a game he played when he was young. Snake is a video game that is available on basic Nokia phones. The player uses the keyboard to control a snake that is trying to pick up the little bits of food represented on the screen. For many South Africans, this simple mobile-phone game was their very first gaming experience (see Chapter 3). The example is typical of many Facebook wall conversations: a status update triggers a jocular response and leads to a short interaction. In this case the interaction is not between a group of friends, but between two male friends, both isiXhosa–English bilinguals in their early twenties. However, being dyadic does not make the interaction private, and the exchange is visible to all their Facebook friends, which, in Thando's case, means over 1,500 people. As in example

(3), we see good-natured banter, teasing and laughter, creating the enjoyment of *plaisir* and affirming community.

(4) Facebook wall interaction (2012)

Status update		
Thando	Just remember how when I was young, I used to take my parent's phone and play Snake!	
Comments		
Khusta	u mean a year ago?:)	
Thando	Lol, zoyibamba le . . . Nc nc nc	'I am going to get this one' [Enacting the voice of the person playing *Snakes* and is hunting for food]
Khusta	hahaha..never!!! u grand mfethu?	'are you cool my friend?'
Thando	Always wena?	'and you?'
Khusta	im good,good brew . . .	
Thando	Buya nini?	'when are you coming back?'
Khusta	maybe next week,maybe never . . .	

The persona we display in such encounters – whether under our real name or using a nick – is usually crafted. Profile pictures are important. Such pictures are carefully selected, and often show us in beautiful locations, surrounded by friends, in various, sometimes deliberately sexy, poses. We are looking at what I called performance-2 in Chapter 6, that is, we present ourselves self-consciously, yet routinely, to others in order to shape the way in which they see us (what Goffman 1969 calls 'impression management'). Although artful and crafted, this is not performance-3. There is usually no 'heightened attention' or sense that something special is happening; rather the displays we put on are a routine part of the social dramaturgy of everyday life.

Managing the way others see us is not only about the images we display, but also about the things we write about; for example, that we bought a beautiful new dress, feel happy, just fell in love, are about to have a baby, are frustrated, had a great night out, just read an amazing book, miss someone and so forth. Although we might share emotions and experiences in such postings, we tend to do so in a stylized way, and what remain hidden are the true 'light and shadow of one's inner life' (Simmel 1949: 256), our actual, personal and intimate feelings. That is, we *play* at intimacy and perform an *illusion* of intimacy, without actually revealing how we feel deep down inside. We present stylized versions of ourselves to others, much as if we were wearing a *mask*.

On Facebook there is a strong sense that one puts on a mask, adopts a 'Facebook persona', follows a certain script and plays for an audience. Moreover, an imperative to 'keep it light' and enjoyable is noticeable among many Facebook users, who describe being 'happy', 'joyful', 'funny', 'bouncy' and 'light-hearted' as desirable character traits when engaging on Facebook. Examples (5)–(7) all come from South Africa, from different users whose circles of friends are dissimilar (in terms of demographics) and non-overlapping. Yet all three pieces of data represent the same sentiment: don't be too serious and if you want to be *emo*, 'emotional', keep it short.

(5) If you have problems, speak about it with somebody, don't put it on your stupid [Facebook] status (interview data 2009; Van Blerk 2010)

(6) [Response to a very unhappy and emotional status update] You need to calm down now and then, review the situation and ask yourself 'what am I doing, why am I doing this' and 'does this really serve as an appropriate Facebook status?' Just a suggestion:) (Facebook 2010)

(7) Can I be emo 4 lyk 2 secndz . . . *
Dis year has had a lot v ups & dwnz.
2012 bekumandi mara new begininz await us . . . 2013 we r on ur door step! #NYECelebrations heleleeee!!! (Facebook 2012)
'2012 was nice but . . . '

Etiquette, masks and conversational co-operation are central to sociability, and the enjoyoment of such interactions remains firmly located within Barthes' *plaisir*. Sociability is about togetherness, 'a union with others', cordiality and good fellowship. The humorous and playful tone of such interactions reflects 'the ludic possibilities of the everyday' (Otsuiji and Pennycook 2010: 246). The notion of the 'everyday' is important here. It is play that is located 'within culture' and that 'comes from culture'. Sociability affirms but does not disrupt the social order. From ludic, conversational engagement it is but one step to Bakhtin's carnival. While sociability is *plaisir*, carnival moves toward *jouissance*.

CARNIVALS: SOCIABILITY SQUARED

Bakhtin's ([1965] 1984) reflections on carnival show parallels with Simmel's discussion of convivial sociability. Carnival, like sociability, is characterized by a 'strong element of play' and 'belongs to the borderline between art and life' (p. 7); laughter is pervasive and interaction is 'free and familiar' (p. 10). We can think of carnival as sociability squared. It is a hyperbolic sociability, exuberant, high-spirited and unruly. The central difference between sociability and carnival is carnival's deliberate disregard for 'etiquette and decency' (p. 10), its intentionally transgressive quality. Consequently, the carnivalesque draws on semiotic resources that are avoided in sociable conversations. There is extensive use of abusive language – profanities and swear words – and a style that appreciates that which is grotesque, exaggerated and excessive.

Laughter is central to Bakhtin's ([1965] 1984: 11, 23) understanding of carnival. Without laughter, there cannot be carnival:

> [C]arnival is the people's second life, organized on the basis of laughter . . . Laughter purifies from dogmatism, from the intolerant and the petrified; it liberates from fanatism and pedantry, from fear and intimidation, from didacticism, naïveté and illusion, from the single meaning, the single level, from sentimentality.

In digital writing, laughter is epitomized in the use of numerous laughing acronyms, many of which have developed into discourse markers (Deumert 2006; on laughter in interaction more generally see Glenn 2003). In their quantitative study of Canadian IM (discussed in Chapter 5), Tagliamonte and Denis (2008: 11) note the presence of 'thousands of uses of *haha*', followed by *LOL* ('laughing out loud') and *hehe*.

Laughter is not exclusive to carnival, and sociability too is infused with laughter. However, the carnival laughter is louder and more boisterous than that of sociability. The different tone of carnival laughter is captured in a range of digital laughing acronyms that contain billingsgate speech in their full form. LMAO, 'laughing my ass off', and LMFAO, 'laughing my fucking ass off', are examples of this. The laughter they express is more exaggerated than that of *haha* or *lol*, and by drawing on taboo regions of the body the acronyms become dysphemisms, offensive in their full form. And just as the writing strategies discussed in the previous chapter are applied cross-linguistically, so laughing acronyms in different languages follow similar strategies and conversational maxims. Writers might localize existing global expressions through word-by-word translations, or create new forms that further dramatize the principle of offense and profanity. Both processes are at play in Afrikaans. On the one hand, there are forms such as LMGA, *lag my gat af*, which is a direct translation of LMAO. On the other hand, there exist novel acronyms that draw on powerful local profanities, especially the derogatory expletive *poes* 'cunt'. The Afrikaans laughter acronyms listed in example (8) can be described as a form of anti-language (Halliday 1978). They establish a mood that violates all that is proper and respectable, and establish an oppositional and transgressive identity. There is also a sense of secrecy surrounding such forms and for those without insider status their meaning remains obscure and the acronym just an incomprehensible string of letters.

(8)
LOPL *laughing out poes* ('cunt') *loud*[6]
LOKL *laughing out kak* ('shit') *loud*
LMGA *lag my gat af* 'laughing my ass off'
LMK *lag my kak* 'laughing me shit-crazy'
LMIMP *lag my in my poes* 'laughing in my cunt'
LMIMMSP *lag my in my ma se poes* 'laughing in my mother's cunt'

LMIMMSFP	*lag my in my ma se fokken poes* 'laughing in my mother's fucking cunt'
LMBIMP	*lag binne in my poes* 'laughing inside my cunt'
LMJ	*lag my jas*, 'laughing myself horny'

These forms can occur in prolonged laughter sequences where they function as interjections, sprinkle everyday interactions and are combined with other forms of laughter (ordinary *ha ha* laughter, *lol* and smiley faces as well as more rowdy *har har* laughter). This is illustrated in examples(9)–(11). In (9), the acronyms are combined with yet another profanity (MSP, *ma se poes*, 'mother's cunt'; used here as an emphasizer). Although this is transgressive and rebellious, the carnival mood ensures that no real offence is taken. Crude profanities are uttered, yet the spirit is humorous.[7]

(9) lmimmsp haha yoh that was msp funny and you still trying to keep your laugh in. Lmbimp (Twitter, male writer, 2013)

(10) Lmbimp I can never go somewhere without having to skarrel ['scrabble'] a lift home (Twitter, female writer, 2013)

(11) Lmimp! Ahhh! ahahahahahahahahahah hahaha! LMK! LMJ! ROTF! LOL! har haaar har! (Twitter, male writer, 2013)

The multimodal nature of digital communication allows the grotesque to move beyond what Bakhtin calls the *language* of the marketplace, loud and unruly, and toward the visual depiction of grotesque *bodies*. The grotesque exaggerates that which is inappropriate, especially bodily parts such as noses, genitals and body mass (Bakhtin [1965] 1984: 315ff.).

Images that circulate on Facebook, often photo-shopped or otherwise modified, frequently foreground the body's carnality and corporeality; its biological profanity and excess (burping, farting, defecating, over-eating); its sexuality (copulating); decline and destruction. Searching on Google Images or scrolling through Facebook walls – many of which are publicly accessible and not privacy-protected – one does not only encounter holiday shots, self-portraits and smiling groups of friends, motivational messages, innocent, child-safe humor and political discussions. There are also images of trees with outgrowths shaped like genitals; meat on the grill whose fatty parts resemble, again, genitals; more images of exposed genitals; a creature that is half man and half snake (Figure 8.1); obscenely overweight, or otherwise repulsive, bodies (Figure 8.2); and people with protruding body parts (Figure 8.3). Such images are typically greeted with laughter and mock-shock, rather than real disgust and offence. Since these are not personal images of known people, the laughter is not directed at particular interlocutors, but is, in Bakhtinian fashion, radical, universal and, essentially, philosophical: it is not 'the drama of an individual body or of a private material way of life; it [is] the drama of the great generic body of the people' (p. 88).

Like any space shaped by human actors, the digital is complex: not merely playful and infused with fellowship and socially acceptable humor (*plaisir*-like, as illustrated in the previous section), it also exhibits the blunt, frank language and imagery of

Figure 8.1 Chimera: man–snake creature (Facebook; posted by a male South African in his thirties, 2011)

Figure 8.2 Sugar-daddy (Facebook, posted by a male South African in his late teens, 2012)

Figure 8.3 Man with big ears (Facebook, localized version posted by a male South African in his twenties, 2014). Translation: 'Who is it, is this the admin of Jix?'(for the complex semantics of *Jix*, see the *Urban Dictionary*)

carnival spaces, intensely bodily, subversive and infused with boisterous laughter. Their punctuated appearances on Facebook newsfeeds and across the internet shape digital interaction in particular ways and remind one of dirty jokes and risqué conversation topics. It is not sociability as discussed above, but also not fully outside of it; after all, carnivals, for all their transgressive qualities, still remain 'within culture'. They are licensed transgression, sociability squared. In this sense the carnivalesque oscillates between *plaisir*, that which is acceptable, and *jouissance*, that which shocks and upsets the social world. And indeed, Barthes (1975: 4) describes the boundary between *plaisir* and *jouissance* as inherently, and necessarily, unstable: 'there is always vascillation – I stumble, I err'.

NOT SAFE FOR WORK: TOWARD *JOUISSANCE*

Some digital spaces push more strongly toward *jouissance*, and among them are those identified as Not Safe for Work (NSFW) in internet parlance; that is, sites that should not be read at work due to their offensive and usually pornographic content. The reason for their 'unsafety' is their lack of monitoring: anything can be posted, nothing will be censored. A well-known NSFW site is the Encyclopedia Dramatica, an intentionally obscene and offensive mock version of Wikipedia that declares itself '[to be] parody and satire . . . to poke fun at everyone and everything'. Articles are often racist and sexist, nothing is sacred, and texts are accompanied by explicit graphics, often of an X-rated nature. The site was launched in 2004 and is modeled on Amrose Bierce's *Devil's Dictionary* (1911), a classic of billingsgate speech.[8] Uncyclopedia is another example of a mock-Wikipedia. Launched in 2005, it contains almost 30,000 satirical, offensive and fake articles. Among them are nonsense entries on 'Dildo Planations', 'Buttworm Barbecue Assball' and 'Anal Twitmas'.[9] Similar sites have been established in other languages, and unlike in the case of Wikipedia proper, the name itself has been localized. For Danish, for example, there is Spademanns Leksikon, a pun that combines the name of the respected encyclopedia Lademanns Leksikon with the slang term for a stupid and clumsy person. The Absurdopedia/Абсурдопедия is available in Russian, the Pekepedia ('Fake-pedia') in Tagalog, and the Frikipaideia/Φρικηπαίδεια ('Horror-pedia') in Greek.

Perhaps the most intriguing of all NSFW's sites is 4chan, an imageboard that was started in 2003. Of particular interest is its random board /b/, which warns novice viewers: 'The stories and information posted here are artistic works of fiction and falsehood. Only a fool would take anything posted here as fact.'[10] What makes 4chan unique is its radical anonymity. Registration is not required and users can simply start posting under the default name 'anonymous'. This site design creates a unique interaction order: if everyone is called 'anonymous', it is impossible to actually figure out who has been talking to whom, and how many people are involved in a conversation. Conventional sociolinguistic analysis is entirely impossible on 4chan. The different voices in any thread could be a monologue of one individual, a conversation of two people, or a multi-layered interaction of several people.

Not only is what is posted not attributable to individuals, it is also fully ephemeral. Although individual users might keep personal archives of discussions they enjoyed, the site itself has no archiving function. The pace is fast, with new posts pushing existing posts to the bottom until they have reached the last page, which self-destructs periodically. Thus, the average post on 4chan spends less than five seconds on the first page and disappears entirely in less than five minutes (Bernstein et al. 2011). To see 4chan in action it is best to log onto its random board and spend some time there. Refreshing frequently allows one to get a sense of the fast movement of posts.

The anonymity and ephemerality of the site shape participation. They help to overcome self-censorship, relax inhibitions and allow for experimentation. No matter how taboo or incendiary, texts cannot be traced back to an author, and boring or lame postings will be bumped off quickly and forgotten. It is the closest we can get to a world where actions don't have consequences, where we don't have to fear lasting ridicule or reprisal. The content on the site is sometimes funny and bizarre, juvenile and ridiculous, frequently repulsive and distasteful. What is detested in 'serious' and 'sober' life is celebrated, giving rise to what is perhaps the internet's 'most prolific semiotic laboratory' (Mendoza 2011: 4), even though, or perhaps because, it resembles at times a 'high-school bathroom stall' (Schwartz 2008) and produces 'X-rated latrinalia' (Phillips 2013). Julian Dibbell (2010) calls 4chan the internet's *id*, its repressed and unruly unconscious. And indeed anything goes: 'trolls, flames, racism, off-topic replies, uncalled for catch-phrases . . . indecipherable text . . . pornography'.[11] Exposed genitals of any shape and size are a staple; there might be images of Hitler with various captions, random photos of broken office chairs and laptops, photos of ordinary people (some ugly, some attractive) and images of cars and guns, interspersed with images of anime characters and space-craft.

The postings on 4chan produce texts of *jouissance*, since much of what is posted discomforts and unsettles 'the reader's historical, cultural, psychological assumptions, the consistency of his takes, values, memories' (Barthes 1975: 14).[12] While 4chan might be one of the darkest corners of the web for some, it is not on the digital fringe, and has been central to internet culture. With 25 million unique visitors per month it certainly lags behind Twitter (190 million), Facebook and YouTube (around 1 billion each), but it is among the top 400 sites in the US, and among the top 800 worldwide.

Barthes' concept of *jouissance* is not only about the *creation* of transgressive texts, but also about the *experience* of transgressive pleasure. It is thus important to take a closer look at the ways in which the texts on 4chan and other NSFW sites are experienced as enjoyable. What kind of enjoyment is this? A good starting point for understanding digital *jouissance* is the online phenomenon of trolling, which is closely associated with the 4chan user base (Phillips 2011, 2013; Coleman 2012). Trolls – a new media variant of the Trickster archetype – take delight in upsetting others, in causing havoc and in creating anarchy. A popular trolling practice is to establish oneself as a regular participant on a particular site and, once accepted, to engage in anti-social behavior and cause disruption, typically by posting inflammatory and offensive messages. The enjoyment trolls experience when people get upset and mad at them is captured by the emic term *lulz*, which relates to the sociable laughter of lol as *jouissance* relates to *plaisir*. The cultural anthropologist Gabriella Coleman (2012: 112) defines lulz as follows:

> The lulz . . . celebrates a form of bliss that revels and celebrates in its own raw power and thus is a form of joy that, for the most part, is divorced from a moral hinge.

Coleman's description of lulz as a form of joy akin to 'bliss' evokes Richard Miller's translation of Barthes' *jouissance* as 'bliss' (Barthes 1975). Yet, as noted by Jane Gallup (2012: 566), 'bliss' is not a good choice; it 'is too airy, too spiritual, too tranquil to carry the edgy, anti-establishment and explicitly sexual connotations of *jouissance*'. Lulz, just like *jouissance*, is a subversive, upsetting, and in its extreme manifestations even sadistic form of humor, which is often sexualized. The practices that give rise to lulz come in degrees. They range from stupid jokes to racial abuse, from exposed breasts to hard-core porn, from frivolous pranks to public attacks on individuals. What unites them is an enjoyment of causing discomfort by violating social norms and expectations.

Although 4chan's radical anonymity is singular, it is not the only transgressive, offensive and ethically suspect site online. In South Africa, Ou Toilet ('old toilet') is based on similar design principles (anonymity and ephemerality), and has become notorious for the sexually charged, colorful and highly abusive insults and malicious gossip users level at named and locally identifiable individuals (Schoon 2012). Another example is the Russian discussion board udaff.com (as well as its provocatively named predecessor fuck.ru). Like 4chan, udaff.com is an anarchic and transgressive space, but unlike 4chan the site requires registration (but encourages pseudonyms, i.e. the use of nicks) and keeps an archive. Those frequenting the site belong to the self-identified subculture of *padonki* (падонки), a deliberate misspelling of *podonok* (подонок), meaning 'jerk', 'riffraff' or 'schmuck'. In addition to willful orthographic distortions – similar to but more aggressive and subversive than what was described in Chapter 7 – *padonki* revel in the use of obscenities, in 'virtuoso and abundant swearing' (Goriunova 2012: 54), and writers refer to themselves as *hujators* ('dicking guys', from хуй, *huī*, 'dick, penis'). Both 4chan and udaff display an aesthetic that regularly ventures into the offensive and distasteful. Their digital carnivals are no longer located *within* culture. They are anarchic, outside of what can be spoken about in public. Their humor is discomforting and typically of the 'laughing at' type, usually directed at women, gay men and ethnic minorities. These texts are produced by a user base that is closely associated with the historical origins of the internet: male, White and young.

When Christopher Poole, aka moot, the founder of 4chan, spoke at a TED talk in 2009 about the 'case for anonymity online', he described 4chan as an 'open place, raw and unfiltered', and argued that it was precisely this openness that supported creativity on the site.[13] The next section further explores masks – anonymity and pseudonymity – as an integral aspect of digital culture, albeit one that seems to be coming increasingly under pressure. While masks can be used to hide who we are as we engage in socially unacceptable behaviors such as trolling, they are also an important aspect of digital creativity. Thus, writers create not only texts, but also imaginary online personae.

THIS IS NOT ME ☺: MASKS AS PLAY AND DECEPTION

In 2006, when *Time Magazine* announced 'the person of the year', they did not pick some great man or great woman, but simply showed on their cover a computer with just one word on the screen 'YOU', followed by the tagline 'Yes, you. You control the information age. Welcome to your world.' The intention was to acknowledge those who contribute user-generated content to the internet, to celebrate a story of 'community and collaboration' (Grossman 2006). For the reader looking at the cover the 'you' implies 'me', a pronoun that is usually interpreted in terms of personal identity, and much recent work on online communication has focused on how people express aspects of their offline identities in digital environments.

In *Language Online* (2013), David Barton and Carmen Lee discuss digital writing under the chapter heading 'This Is Me', and Rodney Jones and Christoph Hafner (2012) title the opening chapter of *Understanding Digital Literacies* 'Mediated Me'. And indeed, with the advent of social networking applications we have witnessed not only the growth of user-generated content, but also the rise of 'me', or what Lee Knuttila (2011) calls the 'personal turn'. That is, online engagement is increasingly rooted in, or at least closely linked to, one's offline identity. I might want to show myself as prettier, smarter and funnier online than I am in real life, but the persona on display is still understood to be 'me' or, at least, the person I would like to be. The real-name policies of Google+ and Facebook are prominent examples of the personal turn. Both sites require users to register with their 'real' name and surname. Failure to do so – although technically possible – is a violation of the terms of service. While Facebook relies on the honesty of users, Google+ closed a number of accounts in 2011 that were identified as being in breach of this policy. This started what has become known as the nymwars: a vigorous debate about the value of online anonymity and pseudonymity (for an overview see Van Zoonen 2013). There are, of course, practical reasons why some platforms prefer people to write under their real name: it facilitates targeted advertising as well as internet security and surveillance. However, the preference for the 'real me' goes deeper and pre-dates 9/11 security concerns and the corporatization of digital platforms. Already in the mid-1980s, the virtual community The Well introduced the slogan YOYOW, 'you own your own words', and insisted on real-name registration: 'As a WELL member, you use your real name. This leads to real conversations and relationships.' Anonymity, on the other hand, is seen as disruptive, dangerous to social relations online, and those who desire it are seen as displaying, in the words of Mark Zuckerberg, the founder of Facebook, a 'lack of integrity' (cited in Kirkpatrick 2010: 36).

Yet the recent valorization of the 'real me/you' stands in contrast to a long tradition of, especially, pseudonymity online. Thus, while 4chan's radical anonymity remains fairly singular, the use of pseudonyms is deeply embedded in the history of the internet. And indeed, regulations notwithstanding, old habits of playful online pseudonymity die hard. One of the people on my Facebook friend list, for example, uses his first name, but combines it with a fictional surname (which reflects his musical passion and artistic persona). Carla Jonsson and Anu Muhonen (2012) describe the imaginary Facebook identity of a Swedish-Finnish teenager whose love for all things Japanese led to the establishment of an entirely fictional Japanese Facebook identity, complete with imagined family, place of education and work.

Some internet researchers have argued that the pseudonyms we establish are ultimately an 'extension of the self', reflecting a practice whereby people merely explore 'different aspects of their "real life" identities' and project a persona that is essentially continuous with the 'real me' (Bechar-Israeli 1995; Jones and Hafner 2012: 79). However, although such 'extensions of the self' are common, what we do online is not always and necessarily an extension of who we are or wish to be; it can also be quite separate from that. It is an important affordance of the digital that it provides escape from the strictures of offline identities that are tied to concrete, physical bodies, and that can be manipulated only within limits. Sherry Turkle (1995: 180; see also Donath 1998), for example, argues that online identities are often creative fabrications that are deliberately and intentionally 'separate from the self'. Thus, as in the notion of metaxis, discussed in Chapter 2, we can be who we are (offline) and be someone quite different online – like any good (or not so good) actor. The concepts of play, imagination and fantasy are important for thinking about nicks and alter personae, as well as about online sociability (see also Hall 1996 for a critical discussion of identity-based approaches to social life more generally).

Using nicks and staying anonymous or pseudonymous allows me to explore identities that are not mine at all, that I do not necessarily desire, but that I might be curious about because they are different from who I am, or because they stimulate my imagination. Consider, for example, the use of nicks reflecting criminal affiliation, which is common practice in some of the dangerous, gang-ridden ghettos of Cape Town. Among teenagers aged between fifteen and seventeen, we find nicks that directly evoke gang culture: *MRS. MAIFA GAL*, *BR!T!$# B!TC#* (the local Hard Living's Gang uses the British flag as a symbol) . . . *"MR:KONVICTED"* . . . and *American Boy* (the Americans are another notorious local gang). An identity-based interpretation would assume that these nicks reflect, in some sense, an aspiration, a wish to belong to the local gang culture. Yet this does not seem to be the case; rather it is a play with danger, with taboo, and a fantasy of strength and toughness (see Coetzee 2012 for further examples). Pseudonymity also allows users to conceal stigmatized aspects of their personality in order to avoid harassment or exclusion. Thus, women might not reveal their gender, or individuals with disabilities might establish an online persona who is able to engage in practices that are otherwise not available to them (Ginsburg 2012).

Pseudonyms are not always playful fantasies; they can also be used to deceive and mislead. Goffman's work is useful here. He can be read not so much as a theorist of deception but as a theorist of how credibility is established within the dramaturgy of social life and the interaction order. Thus, the resources 'whereby we convince others that we are as we appear to be . . . are the same resources confidence tricksters use to deceive people' (Manning 2005: 91). Or in the words of Goffman, audiences accept 'performed cues on faith', and it is this 'sign-accepting tendency' that puts the audience 'in a position to be duped and misled' (1969: 65; see also 1974: 439ff.). The practice of trolling, discussed briefly in the previous section, is an example of this. Internet scams are another genre of deception. Typically, the writer pretends to be someone else (a lawyer, a heiress, an army captain) needing assistance to transfer money out of the country and asking for permission to use the recipient's bank account. In turn, the recipient will be rewarded generously. A successful scam requires quite a bit of skill and writers are not always able to create successful texts, that is, texts that reflect the voice of the assumed persona (Blommaert 2010: 127). In example (12), the inconsistent use of capitals and second-language grammar could be seen as being at odds with the voice of a top official in Syria. However, Jenna Burrell (2008) has suggested that it is also possible to interpret texts of this type as a double-edged performance of authenticity. Thus, Syrians don't necessarily write perfect English, and the mistakes also serve to establish the persona of a second-language learner who is writing under duress.

(12) Internet scam, email (2013)
Attn.
I am Ms. S Ghanem, Ex PO to late Mo'tassim Gaddafi, why am contacting you is because I am looking for a Trustworthy person to help claim US$5.3M Fix deposit a made During one of our trip last year to Southern Africa countries.
Reply if you can and more details will be giving to you.

While we might be fairly immune to the dozens of scams that appear in our inboxes, more elaborate and carefully composed scams can be quite successful. The frequency and success with which such impersonations are practiced online – and even picked up as 'true' by mainstream media, as in the case of the gay-girl-in-Damascus blogging hoax and the entirely fictitious Moldovan soccer sensation Masal Bugduv – show that textual deception can be particularly difficult to detect (Burroughs and Burroughs 2011; Kuntsman and Stein 2011; on the linguistics of lying and deception see Hancock 2007; Toma and Hancock 2012).[14]

Yet not all deception is malicious and, indeed, the ability to lie, that is, to make things up, to create a fantasy world, is essential to being human. The digital includes many examples of creative impersonations that intend not to disrupt, but rather to participate in, and contribute to, social life, albeit under an assumed identity. While a successful deception – as in the case of trolls discussed above – can bring about the enjoyment of lulz or *jouissance*, impersonation can also be *plaisir*.

Eugene Gorny (2009), looking at Russian internet culture in the 1990s, argues that the construction of fictional (or virtual) personae is an important form of online creativity. Virtual personae proper are constructed carefully, over a long time and with considerable detail. An example discussed by Gorny is the persona of Katja Detkina, a popular Russian blogger who was invented and performed by Artemij Lebedev, a blogger as well but male. At other times, writers might 'take over' real-life personalities. Thus, David Gunkel (2011) discusses how the internet magazine *Wired* published an interview with Marshall McLuhan in 1996, more than a decade after his death. The text of the interview was based on correspondence that had taken place in 1996 between Gary Wolf and someone who called himself Marshall McLuhan and had started posting under this name a year earlier. This is the opening paragraph for the article:

> McLuhan (who would have been 85 this year) said he now lives in a beach town in Southern California named 'Parma.' (This town does not exist.) One after another, tiny hints, confirmed by third parties close to McLuhan decades ago, convinced Wolf that if the poster was not McLuhan himself, it was a bot programmed with an eerie command of McLuhan's life and inimitable perspective. [The article then continues with 'the interview', which was conducted online.][15]

Another example was the spoof Twitter account of the embattled former president of the ANC Youth League in South Africa, Julius Malema. Spoof Twitter accounts of politicians have emerged as a common form of political commentary and parody.

The more controversial a political figure, the more likely it is that such accounts proliferate. More than 150 spoof Twitter accounts existed for Julius Malema in September 2012, most of them short-lived without regular tweets. However, one such account was highly active from its establishment in 2009: Julius Sello Malema @ Julius_S_Malema. Unlike the other accounts, which were obvious spoofs, this account seemed to be the voice of 'the real Malema', not only in the political sentiments that were expressed, but also in the style and humor used to convey them. The reason for establishing the account seems to have been a game of make-believe, rather than motivated by political critique. In 2012, the author, who asked to remain anonymous, explained in a newspaper interview:

> I only started it because someone said I was like Malema and I wanted to prove them wrong. I merely started the account and began quoting him verbatim.[16]

The author also commented on the art of successful impersonation:

> It's easier than you think . . . You have to know what to tweet and when . . . My aim wasn't about making fun of anyone, it was more about impersonating.[17]

The account was extremely successful (with more than 200,000 followers in 2012), and created a space for political discussion, banter and social interaction. In this case impersonation was not about troll-like disruption and *jouissance*, but about deeply social *plaisir*. In 2012 the author offered the account to Julius Malema, who took it over and began to write his own tweets (or so the papers say: we can never know for sure, can we?). Examples such as these show that playful impersonations are not necessarily deceitful and disruptive; if they are successful then these personae can contribute to online sociability just as much as those who use their 'real' names and present their 'real' selves.

CONCLUSION: WILD PUBLICS

The sociologist Ray Oldenburg (1989), whose work on third spaces was briefly discussed in the introduction to this chapter, has lamented the disappearance of informal public gathering places in contemporary societies. However, while we might not be hanging out on actual street corners any more, we certainly engage in conversations with others on various digital platforms, and find respite from the daily responsibilities of work, school, intimate relationships and family on Facebook, Twitter, Flickster, YouTube and in many other digital environments. Is it possible that, with the advent of the internet, we are witnessing the constitution and consolidation of a – potentially global and transnational – digital public sphere? A place where all kinds of issues and matters can be debated freely *by those who have access to such technologies*? (The caveat in italics is obviously central: see Chapters 1, 3 and 9; also Lunt and Livingstone 2013.)[18]

Jürgen Habermas' (1962) influential concept of the public sphere has emphasized rational, level-headed and informed discussion of matters of universal, collective importance. Like J.L. Austin (Chapter 5), Habermas discounts ludic language and sees it as less than desirable in the ideal public sphere. His is the ideal of a unified public where language is sincere and truthful, meanings are transparent and clarity of expression is, in principle, achievable. The discussion of Derrida and Bakhtin in Chapters 5 and 6 cast doubt on such a project: unambiguous, unmediated understanding is simply not possible if we accept the principles of *différance* and heteroglossia as fundamental to all language, spoken and written.

Simmel's notion of sociability allows us to think about the public sphere differently. Enjoyment, playfulness, masks and a general 'artistic impulse' are important aspects of the sociable interactions Simmel describes. This is echoed by Michael Gardiner's (2004) discussion of what he calls 'wild publics'. He notes that publics rarely agree on what is important and are not necessarily interested in critical reason. In sociable interaction, art, parody and satire are equally important means of engagement. They not only allow publics to voice their views in creative and playful ways, providing entertainment just as much as debate, but can also allow one to 'smuggle' critique past potential censors (as discussed in Chapter 5).

And sometimes the 'wild' is taken a step further than ordinary sociability: carnivals and *jouissance* are just as much part of human interaction as are fellowship, tact and etiquette, that is, as *plaisir*. Gardiner's notion of the 'grotesque symposium' captures quite aptly some of the data discussed in this chapter. The public sphere is at times a Bakhtinian marketplace, loud, offensive, transgressive and unruly. While some of its carnivalesque practices remain at the borders of what is acceptable, other practices move well beyond and venture into the unacceptable and offensive.

And as we participate in the public sphere we may do so in the guise of many personae, wearing different masks. And not all of these masks are linked to what we, or others, perceive to be 'our identity'. Some of these personae might have a close similarity to the person we consider, or desire, ourselves to be, while others might be at various degrees of distance from it, and even opposed to it. One of the greatest psychological affordances of the digital is that I don't have to be 'me' when I am online. I don't have to voice the same opinions that I tell people offline, yet I can still debate, discuss and contribute. I can explore new ways of thinking, play around with ideas and maybe test the waters for things the offline 'me' finds too daring.

This possibility of anonymity and pseudonymity is a challenge for sociolinguists working in new media contexts. Well-established ways of looking at language and the social world are not possible when we simply don't know whether the text we are analyzing was produced by a young White male or a middle-aged African woman. Under such conditions, sociolinguistic approaches that see language as reflecting speakers'/writers' pre-existing social position and identity are all but impossible. Digital sociolinguistics encourages us to explore other ways of looking at language, and suggests that fantasy, play and creative practice are just as important as conventions, norms and identities.

NOTES

1. The discussion in this paragraph is based on Alexa data (November 2013). However, things can change quickly, and at the time of writing (2013/2014) there was some indication that the popularity of Facebook might be waning, especially among teenagers.
2. http://www.theguardian.com/commentisfree/2013/apr/18/facebook-big-misogyny-problem.
3. I would like to thank Yolandi Klein for this example.
4. There has been popular concern that sometimes such enjoyment goes too far and that people might become 'addicted' to online interaction. In other words, like gambling, online sociability is seen a potential behavioral problem with damaging repercussions for a person's life. These assessments appear to be based on the idea – a version of Derrida's metaphysics of presence (Chapter 2) – that body-to-body interactions are 'healthy' (and it does not appear possible to become addicted to them), but mediated interactions are inherently problematic and should be kept to a minimum (see Kuss and Griffiths 2011).
5. Expectations of sustained conversational engagement feature in several YouTube videos on texting etiquette. A good example is the video 'Don't Flake While I'm Texting You', where flutegirl1995 (from the US) discusses the time-order of texting in detail (http://www.youtube.com/watch?v=ztpyuyxdRyI).
6. This is a local interpretation of a globally circulating acronym which is generally interpreted as a variant of LOL. The *Urban Dictionary* provides a long list of possible glosses, including 'laughing out pretty loud', 'laughing out purple liquid', 'laughing out phucking loud' and 'laughing out pissing loud'. The letter P is thus open to creative interpretation.
7. In spoken language too these profanities can be uttered among friends in a jocular spirit.
8. In April 2014, Encyclopedia Dramatica was hosted at https://encyclopediadramatica.es. The site survives through donations and advertising, including hard-core pornography. Readers access this and other sites mentioned here at their own risk.
9. In 2012, Uncyclopedia was purchased by Wikia, a web-hosting service, which has enforced guidelines of 'acceptable' content. In response to this new policy, a group of administrators created a new site (http://en.uncyclopedia.co), which re-hosts the original site, offensive and sexual content included. The cleaned-up version can be found at http://uncyclopedia.wikia.com.
10. https://boards.4chan.org/b.
11. http://www.4chan.org/rules.
12. However, not all of 4chan is offensive. The site hosts over sixty different imageboards, many of them dedicated to Japanese popular culture. Just over a dozen of the boards are classified NSFW. 4chan is also one of the homes of the political movement Anonymous, which since 2008 has developed into a recognizable form of online/offline political activism and subcultural performance. Like somethingawful.com, 4chan has been central to the early diffusion of a number of popular internet memes (e.g. lolcats).
13. http://www.youtube.com/watch?v=a_1UEAGCo30.
14. The field of text messaging forensics is becoming increasingly important in criminal investigations (Grant 2010). Among the cases which made the news was the murder of Jenny Nicholl by David Hodgson. Essential to the conviction was an analysis of text messages that had been sent from Jenny Nicholl's phone after her disappearance. The writer claimed to be Jenny. However, from stylistic analysis, these messages were later shown to have been written by Hodgson (http://news.bbc.co.uk/2/hi/science/nature/7600769.stm).
15. http://www.wired.com/wired/archive/4.01/channeling.html.
16. http://mg.co.za/article/2012–09–05-malema-claims-his-twitter-handle.

17. http://memeburn.com/2012/09/why-the-man-behind-julius-malemas-largest-spoof-account-just-handed-it-over.
18. Oldenburg, however, rejects the possibility of meaningful virtual sociability and expresses a strong preference for body-to-body interaction (http://www.jwtintelligence.com/2011/01/qa-ray-oldenburg-author-professor-emeritus/#axzz30Xhzac4t).

Chapter 9

Conclusion

This book has considered sociolinguistics, language and sign-making through the lens of the digital. The discussion has drawn on what I have called the ancestors – those who have thought and written about language long before the internet, and whose writings are helpful in interpreting the practices we see in new media contexts (Goffman, Humboldt, Sapir, Derrida, Jakobson, Simmel, Barthes). Just as writing allowed Derrida to grasp the nature of language (always mediated, always characterized by absences, always deferred; see Chapter 6), looking at mobile communication allows sociolinguists to see things about language that are fundamental to the way it works in social life. The digital draws attention to the material aspects of communication and shows intertextuality, heteroglossia, performance and the poetic to be central to meaning-making and sociolinguistic indexicalities. The view of language presented here links with ongoing debates in sociolinguistics on mobility, diversity and creativity, where speakers and writer 'have come to be seen as constantly *refashioning* linguistic and other communicative resources rather than *reproducing* static rules of languages use' (Maybin and Swann 2007: 499; my emphasis).

In the introductory chapter, I argued that media sociolinguistics allows us to study a range of topics that are central to the larger sociolinguistic enterprise: multilingualism and linguistic diversity, language variation and change, style and register, language and identity, language ideologies, language-in-interaction, language and globalization, multimodality and writing. The chapters in this volume have explored various aspects of these topics from different angles by considering different types of data, different theorists and theoretical frameworks. I also identified three broad themes that informed my own thinking about new media: mobility, creativity and inequality. In this conclusion I revisit the three themes.

MOBILITY: PEOPLE, TEXTS AND IDENTITIES

Chapters 2 and 3 foregrounded the mobility of people and devices. Miniaturized mobilities, such as phones, tablets and laptops, allow us a maximum of personal movement while remaining connected to others. The promise of connectivity has been central to the global success of new media. Being able to be away and to stay in touch appears to be a need and desire of people around the world, and makes even those with very limited incomes invest in technology. While movement and the maintenance of connectivity across physical space are important, we should not reduce mobility to geography. Mobility is equally about movement in social spaces (upward/downward social mobility) and across time. Ownership of mobile technologies, for example, can be a symbol of social mobility (Chapters 2 and 3), and the fact that traces of our digital interactions can be archived provides us with temporal mobility; that is, we can revisit past interactions in the here and now (Chapter 2).

Not only are people and devices mobile; texts and linguistic resources are mobile too, and their mobility is greatly facilitated by global communication technologies. The mobility of texts and resources was discussed in detail in Chapters 5 and 6, drawing on the concepts of intertextuality and heteroglossia. One of the hallmarks of semiotic mobility is the appearance of languages, bits of language and other semiotic resources in 'unexpected places' (Pennycook 2012). Being attuned to the mobility of all signs means that we will expect the unexpected and would be surprised if signs, people, languages and objects stayed 'in place'. Bakhtin ([1929/1963] 1984: 202) had already noted that linguistic forms and utterances are 'eternally mobile, eternally fickle', and that meanings can spin out of control as texts move across social, geographical or temporal scales (Chapters 4, 5 and 6). And finally, the digital also enables the mobility, and indeed at times dissolution, of identities (Chapters 6 and 8). Like the stage, new media environments allow us to take on a character and to create imaginary identities.

While physical mobilities and semiotic mobilities are among the celebrated aspects of online spaces, the mobility of identities seems to be under threat. Twenty years ago, the *New Yorker* published a now classic cartoon about digital interaction and the mobility of online identities. The well-known image shows two dogs in front of a computer; one dog says to the other 'On the internet nobody knows that you are a dog.' The cartoon captured the idea that the internet provides a space where we can leave our real-life identities behind, and explore new ways of being. Chapter 8 discussed how anonymity and pseudonymity are increasingly constructed as risk factors, and how, in these debates, 'once-celebrated discourses of multiplicity have been annihilated by constructions of duplicity' (Van Zoonen 2013: 45). The real-name policies of Facebook and Google+ are prominent examples of a move toward more fixed and static online identities.

The 'mobilities paradigm' is coming of age, complete with its own journal, *Mobilities*, and first-generation textbooks. And mobility – as well as immobility – continues to inspire thinking across the social sciences. A recent example is Chris Stroud's (forthcoming) reflection on 'turbulence', that is, disruptive moments that restructure existing systems and, after the proverbial storm, might bring about a *temporary* sense of order. The focus on mobility also brings with it methodological preferences that focus our gaze on *the fleeting* and *the distributed* (Law and Urry 2004; Büscher et al. 2011). In Chapter 6 I argued that it is important to develop a nuanced theoretical understanding of non-routine *moments* in interaction, moments when speakers move into a performance frame, speak or write in an artful and highly reflexive way. The distributed and networked nature of the digital was discussed in

Chapter 2, with reference to Yang's guerilla ethnography and multi-sited ethnography, as well as in Chapters 5 and 7, which looked at the ways in which global audiences engage with, and recontextualize, digital texts and resources.

CREATIVITY: UNCONTROLLABLE

The playful creativity of interactive digital communication has been a theme throughout the book and is deeply embedded in the history of the medium. The early days of the internet were characterized not only by the dominance of a particular language (English), but also by a particular cultural slant, namely, the subculture of programmers and system designers. A central characteristic of this subculture was (and is) a subversive and broadly playful mood (in the sense of 'playing around' rather than 'playing a game' with rules and well-defined player roles). This is reflected, for example, in the introduction to the Jargon File, a popular manifesto of computer programmer slang that was started in 1975, and is updated periodically:

> Hackers[1], as a rule, love wordplay and are very conscious and inventive in their use of language. These traits seem to be common in young children, but the conformity-enforcing machine we are pleased to call an educational system bludgeons them out of most of us before adolescence. (Jargon File 2004)

Here, digital writing is positioned in direct opposition to a restrictive education system and the 'red pen of the teacher' mentioned at the end of Chapter 7. And consequently, creativity and play are favored and valued, and conformity is disfavored. The broadly ludic mood is also visible in the types of interactions people engage in online – they play games, joke, flirt or just hang out with one another – as well as in the language and multimodal imagery they use (Deumert 2014; see also Chapters 5–7).

The playful creativity we see in digital spaces has been described as everyday or vernacular creativity and is, at times, distinguished from the more serious forms of creativity associated with high art (see, for example, Burgess 2006; Thurlow 2012). While it is debatable whether we can – or even want to – draw a line between everyday and high-art creativity, it is important for sociolinguists to look carefully at the kinds of creativities we see in new media environments in order to understand the possibilities for novelty as well as the constraints within which writers/speakers operate. Online creativity – like everyday creativity more generally – is closely linked to sociability, playful and artful interactions that affirm people's social connections and bring about enjoyment (of *plaisir* or *jouissance*; Chapter 8). The link between creativity, sociability and enjoyment is at the heart of David Gauntlett's (2011: 76) book *Making is Connecting*. He argues that when we create something, not only do we connect pre-existing signs or materials in novel ways, but as we do so we also share them with others, establish and reaffirm social connections, and in doing so experience feelings of joy.

Creativity is not only an enjoyable social practice, but also fundamentally illimitable, boundless and ultimately uncontrollable. This point was emphasized by Sapir in his thought experiment about Two Crows, which I discussed in Chapter 2. It can also be illustrated by reconsidering Noam Chomsky's ([1957] 2002: 15) classic discussion of *Colorless green ideas sleep furiously* as an unusual but nevertheless grammatical sentence. Although Chomsky's writings brought the idea of human creativity back into linguistics, he conceives of it solely as 'creative action within the system of a framework of rules' (from his speech 'Language and Freedom', 1970; cited in Smith 2004: 184). Thus, while speakers might create an infinite number of sentences following the syntax of *Colorless green ideas sleep furiously* (*Tired old linguists write continuously, Cute bouncing dogs run happily*, and so forth), they will – according to Chomsky – not break the rules of syntax and rearrange words in unprecedented ways: **Furiously sleep ideas green colorless* is an impossible sentence for Chomsky and would not be uttered by a capable speaker. Yet the syntax of that seemingly ungrammatical sentence is not altogether unfamiliar and might well occur in poetry: *softly moves the sun, gentle, warming.* The sentence also inspired at least one poet: the Australian Clive James used it as the opening line for the poem 'A Line and a Theme from Noam Chomsky' in his anthology *Other Passports: Poems 1958–1985* (published 1986).

The examples in the preceding chapters too have shown that language is both constraint and freedom. Traditionally, sociolinguists have focused their attentions on constraints and norms, the habitual patterns that give rise to distinct and fairly well-defined social practices (Bakhtin's centripetal forces). However, although we usually speak and write in accordance with existing conventions, any of these conventions can be broken, subverted or changed (Bakhtin's centrifugal forces; Chapter 6). To return to the conclusion of Chapter 7, although mobile communication throws the unpredictability and openness of language into relief, I suggest that *all language is liquid*, not only language online. There are innumerable ways in which to speak, to write, to signify. Signs are never closed, but can always be manipulated, twisted and changed. This is also a central point in Derrida's and Butler's work on citationality, signification and performativity (Chapters 5 and 6). Moreover, the meanings that audiences assign to an utterance in new contexts can never be predicted or controlled; they too are *liquid*, capable of multiplying and spinning out of control (see the examples of 'Xhosa Lesson 2' in Chapter 5 and my episode as engen in Chapter 6).

INEQUALITY: PERSISTENT BUT CHANGING

Chapter 3 focused on the material conditions of digital access and provided an overview of the digital divide. My own experience with digital media has mostly been of the Southern type. Although South Africa is fairly affluent within the global South, its realities are profoundly different from those of the North, and high data costs and low bandwidth shape the experience of being online. This is most noticeable with regard to multimodal material. Watching a 3-minute YouTube video can easily take 10 minutes or more as the video loads, stops, plays for a few seconds, stops again and so forth. Yet I am also among the privileged. I have ADSL access at home, a laptop, a tablet, a smartphone, an office computer and free, fast internet at work. For the majority of people in the global South this type of connectivity remains a luxury.

In South Africa the percentage of those without any internet access remains at about two-thirds of the total population. This is high compared to Denmark, where less than 10 percent of the population is unconnected, but certainly better than Pakistan, where more than 90 percent remain without internet access (ITU 2013). The issue of inequality – who has access, and what kind of access – matters whenever we write about mobile communication. In Chapter 8 I suggested – like many before me – that we might be witnessing the emergence of a new, 'wild' public sphere in public and semi-public online environments. But there is a problem with this argument, namely the fact that the public sphere must be accessible to everyone; otherwise it is not a *public* sphere. Yet if the internet is currently only available to less than half of the world's population, how can the digital constitute a new *public* sphere? Similarly, the South African media scholar Marion Walton (2011: 48) has asked provocatively with respect to Africa: 'Can the internet really be counted as a "commons" on a continent where only 10% of the population access on-line media?' Considering who is excluded, whose voices are not being heard and whose voices are the loudest is an important question for media scholars. Moreover, for many in the global South, internet access is via mobile feature phones or through public access venues such as internet cafés or libraries. I argued in Chapter 3 that the materiality of access affects online creativity and content production, and that the read/write-web is experienced quite differently across the world. For many people, playing around with typography is all the multimodality they can afford. The internet is unlikely to transform existing inequalities in global content production any time soon, and much of what is produced still comes out of Euro-America (see also Wall 2009). However, things are changing slowly but steadily, and we can expect to hear more diverse voices online in the years to come.

And this brings me to a final point: the question of voice, which was discussed from a Bakhtinian perspective in Chapter 6. Bakhtin's concept of voice focuses on the diversity of social voices that are available to us as linguistic resources. They allow us to shape and create our own voice, 'the speaking personality, the speaking consciousness' (Bakhtin [1934/1935] 1981). Dell Hymes (1996) provides a slightly different perspective on voice and foregrounds issues of inequality more strongly. He emphasizes the fact that not all voices are equally heard. Having access to technology is a prerequisite for making one's voice heard in the digital space, but access alone does not solve the question of linguistic inequality. As in offline contexts, minority language speakers often have to shift to a majority language in order to 'be heard', and certain texts and linguistic forms struggle to move to scale levels beyond the local (Blommaert 2008; Blommaert and Rampton 2011). Thus, *mic* for *miss* or *xo* for *sure* might indicate stylish writing in South Africa, but are unlikely to catch on globally as – at this stage – they remain largely invisible to global audiences. Variants popular in the global North, on the other hand, have spread to the South (Chapters 6 and 7).

Future sociolinguistic work needs to bring inequality – and thus issues of power – more strongly into the study of new technologies. Inequalities affect the types of creativities users can engage in, the types of mobilities they can experience, and their ability to take advantage of the networked capabilities of the internet to make themselves heard beyond the local.

NOTE

1. The use of 'hacker' as a negative term to describe someone who breaks into a computer's security system differs from its use and positive connotation as a self-identifier for programmers and system designers.

References

All web resources were live as of April 2014.

Agha, A. 2003. The Social Life of Cultural Value. *Language and Communication* 23: 231–73.

Agha, A. 2005. Voice, Footing, Enregisterment. *Journal of Linguistic Anthropology* 15: 38–59.

Agha, A. 2011. Meet Mediatization. *Language and Communication* 31: 163–70.

Aitchinson, J. and Lewis, D.M. 2003. *New Media Language*. London: Routledge.

Akinnaso, F.N. 1982. On the Differences between Spoken and Written Language. *Language and Speech* 25: 97–125.

Alexander, D.M. 1929. Why Not 'U' for 'You'? *American Speech* 5: 24–5.

Alim, S.H. 2011. Global ILL-Literacies: Hip-Hop Cultures, Youth Identities, and the Politics of Literacy. *Review of Research in Education* 35: 120–46.

Allen, G. 2011. *Intertextuality*. Second edition. London: Routledge.

Althusser, L. 1971. Ideology and Ideological State Apparatuses. In *Lenin and Philosophy and other Essays*, trans. B. Brewster, pp. 121–76. New York: Monthly Review Press.

Anderson, B. 1991. *Imagined Communities: Reflections on the Origin and Spread of Nationalism*. Revised and extended edition. London: Verso.

Androutsopoulos, J. 2000. Non-Standard Spellings in Media Texts: The Case of German Fanzines. *Journal of Sociolinguistics* 4: 514–33.

Androutsopoulos, J. 2006. Multilingualism, Diaspora, and the Internet: Codes and Identities on German-Based Diaspora Websites. *Journal of Sociolinguistics* 10: 429–50.

Androutsopoulos, J. 2008. Discourse-Centred Online Ethnography. *Language@Internet* 5. Available at http://www.languageatinternet.de.

Androutsopoulos, J. 2010. Localising the Global on the Participatory Web. In *Language and Globalization*, ed. N. Coupland, pp. 203–31. Oxford: Wiley-Blackwell.

Androutsopoulos, J. 2011. From Variation to Heteroglossia in the Study of Computer-Mediated Discourse. In *Digital Discourse: Language in the New Media*, ed. C. Thurlow and K. Mroczek, pp. 277–97. Oxford/New York: Oxford University Press.

Androutsopoulos, J. 2013. Networked Multilingualism: Some Language Practices on Facebook and Their Implications. *International Journal of Bilingualism* Online preprint.

Anis, J. 2007. Neography: Unconventional Spelling in French SMS Messages. In *The Multilingual Internet: Language, Culture and Communication Online*, ed. B. Danet and S. Herring, pp. 87–115.Oxford/New York: Oxford University Press.

Apollinaire, G. 1918. *Calligrammes: Poèmes de la paix et de la guerre 1913–1916*. Paris: Mercure de France.

Appadurai, A. 1996. *Modernity at Large: Cultural Dimensions of Globalization*. Minneapolis/London: University of Minnesota Press.

Archambault, J. S. 2012. 'Travelling while sitting down': Mobile Phones, Mobility and the Communication Landscape in Inhambane, Mozambique. *Africa* 82: 393–412.

Archer, M. 2012. *The Reflexive Imperative in Late Modernity*. Cambridge: Cambridge University Press.
Auer, P. 2000. On Line-Syntax – oder: Was es bedeuten könnte, die Zeitlichkeit der mündlichen Sprache ernst zu nehmen. *Sprache und Literatur* 85: 43–56.
Austin, J.L. 1962. *How to Do Things With Words*. Second edition. Ed. J.O. Urmson and M. Sbisà. Cambridge, MA: Harvard University Press.
Back, M. and Zepeda, M. 2013. Performing and Positioning Orthography in Peruvian CMC. *Journal of Computer-Mediated Communication* 18: 119–35.
Bailey, B. 2012. Heteroglossia. In *The Routledge Handbook of Multilingualism*, ed. M. Martin-Jones, A. Blackledge and A. Creese, pp. 499–507. London: Routledge.
Bakhtin, M.M. [1929/1963] 1984. *Problems of Dostoevsky's Poetics*. Ed. and trans. C. Emerson. Minneapolis: University of Minnesota Press.
Bakhtin, M.M. [1934/1935] 1981. Discourse in the Novel. In *The Dialogic Imagination*, trans. C. Emerson and M. Holquist, pp. 259–422. Austin: University of Texas Press.
Bakhtin, M.M. [1952/1953] 1986. The Problem of Speech Genres. In *Speech Genres and Other Late Essays*, trans. V.W. McGee, pp. 60–102. Austin: University of Texas Press.
Bakhtin, M.M. [1965] 1984. *Rabelais and His World*. Trans. H. Iswolsky. Bloomington/Indianapolis: Indiana University Press.
Bamberg, M. and Georgakopoulou, A. 2008. Small Stories as a New Perspective in Narrative and Identity Analysis. *Text & Talk* 28: 377–96.
Barasa, S.N. 2010. *Language, Mobile Phones and Internet: A Study of SMS Texting, Email, IM and SNS Chats in Computer Mediated Communication (CMC) in Kenya*. Unpublished PhD dissertation, University of Leiden.
Barendregt, B. 2012. Diverse Digital Worlds. In *Digital Anthropology*, ed. H.A. Horst and D. Miller, pp. 203–24. London/New York: Berg.
Bargh, J.A. and McKenna, K.Y.A. 2004. The Internet and Social Life. *Annual Review of Psychology* 55: 573–90.
Baron, N. 2008. *Always On: Language in an Online and Mobile World*. Oxford: Oxford University Press.
Barthes, R. [1967] 1977. The Death of the Author. In *Image, Music, Text*, trans. S. Heath, pp. 142–8. New York: Hill and Wang.
Barthes, R. 1975. *The Pleasure of the Text*. Trans. R. Miller. New York: Hill and Wang.
Barthes, R. [1977] 2002. *A Lover's Discourse: Fragments*. Trans. R. Howard. London: Vintage.
Barton, D. 2007. *Literacy: An Introduction to the Ecology of Written Language*. Oxford: Blackwell.
Barton, D. and Lee, C. 2013. *Language Online: Investigating Digital Texts and Practices*. London: Routledge.
Basso, K. 1974. The Ethnography of Writing. In *Explorations in the Ethnography of Speaking*, ed. R. Bauman and J. Sherzer, second edition, pp. 425–32. Cambridge: Cambridge University Press.
Baudrillard, J. 1983. *Simulations*. Trans. P. Foss, P. Patton and P. Beitchman. New York: Semiotext(e).
Bauman, R. 1972. The La Have Island General Store: Sociability and Verbal Art in a Nova Scotia Community. *Journal of American Folklore* 85: 330–43.
Bauman, R. 1977. *Verbal Art as Performance*. Prospect Heights, IL: Waveland.
Bauman, R. 2004. *A World of Others' Words: Cross-Cultural Perspectives on Intertextuality*. Oxford: Blackwell.
Bauman, R. 2011. Commentary: Foundations in Performance. *Journal of Sociolinguistics* 15: 707–20.
Bauman, R. and Briggs, C. 1990. Poetics and Performance as Critical Perspectives on Language and Social Life. *Annual Review of Anthropology* 19: 59–88.
Bauman, Z. 2012. *Liquid Modernity*. Cambridge: Polity.

Baym, N.K. 1995. The Performance of Humour in Computer-Mediated Communication. *Journal of Computer-Mediated Communication* 1. Available at http://jcmc.indiana.edu/vol1/issue2/baym.html.

Baym, N.K. 2010. *Personal Connections in the Digital Age*. Cambridge: Polity.

Baynham, M. and Prinsloo, M. (eds.) 2009. *The Future of Literacy Studies*. Basingstoke: Palgrave Macmillan.

Bechar-Israeli, H. 1995. From <Bonehead> To <cLoNehEAd>: Nicknames, Play, and Identity in Internet Relay Chat. *Journal of Computer-Mediated Communication* 1. Available at http://jcmc.indiana.edu/vol1/issue2/bechar.html.

Becker, A.L. 1995. *Beyond Translation: Essays Toward a Modern Philology*. Ann Arbor: University of Michigan Press.

Beeman, W.O. 1993. The Anthropology of Theater and Spectacle. *Annual Review of Anthropology* 22: 369–93.

Bell, A. 1984. Language Style as Audience Design. *Language in Society* 13: 145–204.

Bell, A. 2001. Back in Style: Reworking Audience Design. In *Style and Sociolinguistic Variation*, ed. P. Eckert and John R. Rickford, pp. 139–69. Cambridge: Cambridge University Press.

Bell, G. 2006. No More SMS from Jesus: Ubicomp, Religion and Techno-Spiritual Practices. In *UbiComp 2006: Ubiquitous Computing*, 8th International Conference, September 17–21, ed. P. Dourish and A. Friday, pp. 141–58. Berlin/Heidelberg: Springer.

Bernstein, M.S., Monroy-Hernández, A., Harry, D., André, P., Panovich, K. and Vargas, G. 2011. 4chan and /b/: An Analysis of Anonymity and Ephemerality in a Large Online Community. In *ICWSM: International Conference on Weblogs and Social Media 2011*. Available at http://hci.stanford.edu/msb.

Besnier, N. 1995. *Literacy, Emotion and Authority: Reading and Writing on a Polynesian Atoll*. Cambridge: Cambridge University Press.

Best, S. and Kellner, D. 1999. Debord, Cybersituations, and the Interactive Spectacle. *SubStance* 28: 129–56.

Biber, D. 1991. *Variation across Speech and Writing*. Cambridge: Cambridge University Press.

Bickerton, D. 1981. *Roots of Language*. Ann Arbor: Karoma.

Bierce, A. 1911. *The Devil's Dictionary*. Cleveland/New York: World's Publishing Company.

Bieswanger, M. 2006. 2 abbrevi8 or not 2 abbrevi8: A Contrastive Analysis of Different Shortening Strategies in English and German Text Messages. *Texas Linguistics Forum* 50. Available at http://studentorgs.utexas.edu/salsa/proceedings/2006/Bieswanger.pdf.

Billings, A.C. 2008. *Olympic Media: Inside the Biggest Show on Television*. London: Routledge.

Blackledge, A. and Creese, A. 2010. *Multilingualism: A Critical Perspective*. London/New York: Continuum.

Blommaert, J. 2005. *Discourse: A Critical Introduction*. Cambridge: Cambridge University Press.

Blommaert, J. 2008. *Grassroots Literacy: Writing, Identity and Voice in Central Africa*. London: Routledge.

Blommaert, J. 2010. *The Sociolinguistics of Globalization*. Cambridge: Cambridge University Press.

Blommaert, J. 2012a. Lookalike Language. *English Today* 28: 62–4.

Blommaert, J. 2012b. Supervernaculars and Their Dialects. *Dutch Journal of Applied Linguistics* 1: 1–14.

Blommaert, J. 2013. *Ethnography, Superdiversity and Linguistic Landscapes: Chronicles of Complexity*. Clevedon: Multilingual Matters.

Blommaert, J., Collins, J. and Slembrouck, S. 2005. Spaces of Multilingualism. *Language & Communication* 25: 197–216.

Blommaert, J. and Rampton, B. 2011. Language and Superdiversity. *Diversities* 13: 1–22.

Boal, A. 1995. *The Rainbow of Desire: The Boal Method of Theatre and Therapy*. London: Routledge.

Boellstorff, T. 2008. *Coming of Age in Second Life: An Anthropologist Explores the Virtual Human*. Princeton: Princeton University Press.

Bolinger, D.L. 1946. Visual Morphemes. *Language* 22: 333–40.

Bolter, J.A. and Grusin, R. 1999. *Remediation: Understanding New Media*. Cambridge, MA: MIT Press.

Bortzmeyer, S. 2012. Multilingualism and the Internet's Standardisation. In *NET.LANG: Towards the Multilingual Cyberspace*, ed. L. Vannini and H. Le Crosnier, pp. 105–18. Caen: C&F éditions.

Bourdieu, P. 1991. *Language and Symbolic Power*. Harvard: Harvard University Press.

boyd, d. 2002. *Faceted Id/entity: Managing Representation in a Digital World*. Unpublished MA thesis, MIT.

boyd, d. 2008. Why Youth ♥ Social Network Sites: The Role of Networked Publics in Teenage Social Life. In *Youth, Identity, and Digital Media*, ed. David Buckingham, pp. 119–42. Cambridge, MA: MIT Press.

boyd, d. 2010. Social Network sites as Networked Publics: Affordances, Dynamics, and Implications. In *A Networked Self: Identity, Community, and Culture on Social Network Sites*, ed. Z. Papacharissi, pp. 39–58. New York: Routledge.

boyd, d. and Crawford, K. 2012. Critical Questions for Big Data: Provocations for a Cultural, Technological, and Scholarly Phenomenon. *Information, Communication, & Society* 15: 662–79.

boyd, d. and Ellison, N.B. 2007. Social Network Sites: Definition, History and Scholarship. *Journal of Computer-Mediated Communication* 13. Available at http://jcmc.indiana.edu/vol13/issue1/boyd.ellison.html.

Brandist, C. and Lähteenmäki, M. 2011. Early Soviet Linguistics and Mikhail Bakhtin's Essays on the Novel of the 1930s. In *Politics and the Theory of Language in the USSR 1917–1938: The Birth of Sociological Linguistics*, ed. C. Brandist and K. Chown, pp. 69–87. London: Anthem Press.

Brodersen, A., Scellato, S. and Wattenhofer, M. 2012. YouTube Around the World: Geographic Popularity of Videos. In *WWW: Proceedings of the 21st International Conference on the World Wide Web*, pp. 241–50. New York: ACM.

Brown, C. and Czerniewicz, L. 2010. Debunking the 'Digital Native': Beyond Digital Apartheid, Towards Digital Democracy. *Journal of Computer-Assisted Learning* 26: 357–69.

Brown, K., Campell, S.W. and Ling, R. 2011. Mobile Phones Bridging the Digital Divide for Teens in the US? *Future Internet* 3: 144–58.

Bruns, A. 2008. *Blogs, Wikipedia, Second Life, and Beyond: From Production to Produsage*. New York: Peter Lang.

Bucholtz, M. and Hall, K. 2004. Theorizing Identity in Language and Sexuality Research. *Language in Society* 33: 501–47.

Bucholtz, M. and Hall, K. 2008. All of the Above: New Coalitions in Sociocultural Linguistics. *Journal of Sociolinguistics* 12: 401–31.

Burgess, J.E. 2006. Hearing Ordinary Voices: Cultural Studies, Vernacular Creativity and Digital Storytelling. *Journal of Media and Cultural Studies* 20: 201–14.

Burrell, J. 2008. Problematic Empowerment: West African Internet Scams as Strategic Misrepresentation. *Information Technologies and International Development* 4: 15–30.

Burrell, J. 2009. Could Connectivity Replace Mobility? An Analysis of Internet Café Use Patterns in Accra, Ghana. In *Mobile Phones: The New Talking Drums of Everyday Africa*, ed. M. De Bruijn, F. Nyamnjoh and I. Brinkman, pp. 151–70. Bamenda/Leiden: Langaa/African Studies Centre.

Burrell, J. 2010. Evaluating Shared Access: Social Equality and the Circulation of Mobile Phones in Rural Uganda. *Journal of Computer-Mediated Communication* 15: 230–50.

Burroughs, B. and Burroughs, W.J. 2011. The Masal Bugduv Hoax: Football Blogging and Journalistic Authority. *New Media & Society* 14: 476–91.
Büscher, M., Urry, J. and Witchger, K. (eds.) 2011. *Mobile Methods*. London: Routledge.
Butler, J. 1993. *Bodies that Matter: On the Discursive Limits of 'Sex'*. London: Routledge.
Butler, J. 1997. *Excitable Speech: A Politics of the Performative*. London: Routledge.
Butler, J. 1999. *Gender Trouble: Feminism and the Subversion of Identity*. Second edition. London: Routledge.
Butler, J. 2011. Response: Performative Reflections on Love and Commitment. *Women's Studies Quarterly* 39: 236–9.
Cameron, D. and Kulick, D. 2003. *Language and Sexuality*. Cambridge: Cambridge University Press.
Carter, R. 2004. *Language and Creativity: The Art of Common Talk*. London: Routledge.
Castells, M. 1996. *The Rise of the Network Society. The Information Age: Economy, Society and Culture. Vol. I*. Cambridge, MA: MIT Press.
Chafe, W. and Tannen, D. 1987. The Relation between Written and Spoken Language. *Annual Review of Anthropology* 16: 386–407.
Chant, S. and Mcllwaine, C. 2009. *Geographies of Development in the 21st Century: An Introduction to the Global South*. Cheltenham: Edward Elgar.
Chomsky, N. [1957] 2002. *Syntactic Structures*. Berlin: De Gruyter.
Chomsky, N. 1965. *Cartesian Linguistics*. New York: Harper and Row.
Coetzee, F. 2012. *The Multilingual Literacy Practices of Residents Living in a Coloured, Afrikaans-Dominant Neighbourhood in Cape Town: A Sociolinguistic Study*. Unpublished MA thesis, University of Cape Town.
Cohen, N. 2007. Some Errors Defy Fixes: A Typo in Wikipedia's Logo Fractures the Sanskrit. *New York Times*. Available at http://www.nytimes.com/2007/06/25/technology/25wikipedia.html?_r=0.
Coleman, G.E. 2012. Phreaks, Hackers and Trolls: The Politics of Transgression and Spectacle. In *The Social Media Reader*, ed. M. Mandiberg, pp. 99–119. New York: New York University Press.
Comaroff, J. and Comaroff, J. 1991. *Of Revelation and Revolution: Christianity, Colonialism and Consciousness in South Africa, Vol. 1*. Chicago: University of Chicago Press.
Comaroff, J. and Comaroff, J. 2012. Theory from the South: Or, How Euro-America is Evolving Toward Africa. *Anthropological Forum: A Journal of Social Anthropology and Comparative Sociology* 22: 113–31.
Connell, R. 2007. *Southern Theory: The Global Dynamics of Knowledge in the Social Science*. Cambridge: Polity.
Coupland, N. 2007. *Style: Language Variation and Identity*. Cambridge: Cambridge University Press.
Coupland, N. 2009. The Mediated Performance of Vernaculars. *Journal of English Linguistics* 37: 284–300.
Creeber, G. and Martin, R. 2008. *Digital Culture: Understanding New Media*. Maidenhead: Open University Press.
Cresswell, T. 2012a. *On the Move: Mobility in the Modern Western World*. London: Routledge.
Cresswell, T. 2012b. Mobilities II: Still. *Progress in Human Geography* 36: 645–53.
Crystal, D. 1998. *Language Play*. London: Penguin.
Crystal, D. 2006. *Language and the Internet*. Second edition. Cambridge: Cambridge University Press.
Crystal, D. 2008. *Txtng: The G8 Db8*. Oxford: Oxford University Press.
Crystal, D. 2011. *Internet Linguistics*. London: Routledge.
Curran, J. and Park, M.-J. (eds.) 2000. *De-Westernizing Media Studies*. London: Routledge.
D'Arcy, A. and Young, T.M. 2012. Ethics and Social Media: Implications for Sociolinguistics in the Networked Public. *Journal of Sociolinguistics* 16: 532–46.

D'Haenens, L. and Ogan, C. 2013. Internet-Using Children and Digital Inequality: A Comparison between Majority and Minority Europeans. *Communications* 38: 41–60.

Danby, S.J., Butler, C. and Emmison, M. 2009. When 'Listeners Can't Talk': Comparing Active Listening in Opening Sequences of Telephone and Online Counselling. *Australian Journal of Communication* 36: 91–113.

Danet, B. 2001. *Cyberpl@y: Communication Online*. Oxford/New York: Berg.

Danet, B. and Herring, S. C. (eds.) 2007. *The Multilingual Internet: Language, Culture and Communication Online*. Oxford: Oxford University Press.

Dant, T. 2004. The Driver-Car. *Theory Culture Society* 21: 61–79.

Darley, A. 2000. *Visual Digital Culture: Surface Play and Spectacle in New Media Genres*. London: Routledge.

Dawkins, R. 1976. *The Selfish Gene*. Oxford/New York: Oxford University Press.

De Bruijn, M., Nyamnjoh, F. and Brinkman, I. 2009. *Mobile Phones: The New Talking Drums of Everyday Africa*. Bamenda/Leiden: Langaa/African Studies Centre.

De Lanerolle, I. 2012. *The New Wave*. Available at http://www.networksociety.co.za.

De Saussure, F. [1916] 2013. *Course in General Linguistics*. Trans. and annotated R. Harris. London/New York: Bloomsbury.

Debord, G. [1967] 1994. *The Society of the Spectacle*. Trans. D. Nicholson-Smith. New York: Zone Books.

Del-Teso-Craviotto, M. 2008. Gender and Sexual Identity Authentication in Language Use: The Case of Chat Rooms. *Discourse Studies* 10: 251–70.

Deleuze, G. [1977] 2002. The Actual and the Virtual. In *Dialogues II*, trans. E.R. Albert, pp. 112–15. London/New York: Continuum.

Deleuze, G. and Guattari, F. 1987. *A Thousand Plateaus: Capitalism and Schizophrenia*. London/New York: Continuum.

Derrida, J. [1968] 1982. Differérance. In *Margins of Philosophy*, trans. A. Bass, pp. 1–28. Chicago: University of Chicago Press.

Derrida, J. [1972] 1988. Signature Event Context. In *Limited Inc.*, trans. S. Weber and J. Mehlman, pp. 1–24. Evanston: Northwestern University Press.

Derrida, J. 1974. *Of Grammatology*. Trans. G.C. Spivak. Baltimore/London: Johns Hopkins University Press.

Derrida, J. 1998. *Archive Fever: A Freudian Impression*. Trans. E. Prenowitz. Chicago/London: University of Chicago Press.

Deumert, A. 2006. Semantic Change, the Internet, and Text Messaging. In *The Encyclopaedia of Language and Linguistics. Vol. XI*, ed. K. Brown et al., pp. 121–4. Oxford: Pergamon Press.

Deumert, A. 2013. Language, Culture and Society. In *The Oxford Handbook of the History of Linguistics*, ed. K. Allan, pp. 655–75. Oxford: Oxford University Press.

Deumert, A. 2014. The Performance of a Ludic Self on Social Network(ing) Sites. In *The Language of Social Media: Communication and Community on the Internet*, ed. P. Seargeant and C. Tagg, pp. 23–45. Basingstoke: Palgrave/Macmillan.

Deumert, A. forthcoming a. Sites of Struggle and Possibility in Cyberspace: Wikipedia and Facebook in Africa. In *The Media and Sociolinguistic Change*, ed. J. Androutsopoulos. Berlin: De Gruyter.

Deumert, A. forthcoming b. KLK CC: Conformity and Transgression on an On-Line Educational Site. In *Language, Literacy and Diversity*, ed. C. Stroud and M. Prinsloo. London: Routledge.

Deumert, A. and Lexander, K.V. 2013. Texting Africa: Writing as Performance. *Journal of Sociolinguistics* 17: 522–46.

Deumert, A. and Masinyana, O.S. 2008. Mobile Language Choices: The Use of English and isiXhosa in Text Messages (SMS), Evidence from a Bilingual South African Sample. *English World-Wide* 29: 117–48.

Deumert, A. and Mesthrie, R. 2012. Contact in the African Area: A Southern African Perspective. In The *Oxford Handbook of the History of English*, ed. T. Nevalainen and E.C. Traugott, pp. 549–59. Oxford/New York: Oxford University Press.
Dibbell, J. 1993. A Rape in Cyberspace, or How an Evil Clown, a Haitian Trickster Spirit, Two Wizards, and a Cast of Dozens Turned a Database into a Society. *Village Voice*, 21 December.
Dibbell, J. 2010. Radical Opacity. *MIT Technology Review*, 23 August 2010.
Donath, J. 1998. Identities and Deception in the Virtual Community. In *Communities in Cyberspace*, ed. M. Smith and P. Kollock, pp. 29–59. London: Routledge.
Dong, J., Du, C.X., Juffermans, K., Li, J., Varis, P. and Wang, X. 2012. Chinese in a Superdiverse World. In *Papers of the Anéla 2012 Applied Linguistics Conference*, ed. N. de Jong, K. Juffermans, M. Keijzer and L. Rasier, pp. 349–66. Delft: Eburon.
Donner, J. 2007. The Rules of Beeping: Exchanging Messages via Intentional Missed Calls on Mobile Phones. *Journal of Computer-Mediated Communication* 13: 1–22.
Dorleijn, M. and Nortier, J. 2009. Code-Switching and the Internet. In *The Cambridge Handbook of Linguistic Code-Switching*, ed. B.E. Bullock and A.J. Toribio, pp. 127–41. Cambridge: Cambridge University Press.
Dresner, E. and Herring, S.C. 2012. Emoticons and Illocutionary Force. In *Philosophical Dialogue: Writings in Honor of Marcelo Dascal*, ed. D. Riesenfel and G. Scarafile, pp. 59–70. London: College Publication.
Eckert, P. 2008. Variation and the Indexical Field. *Journal of Sociolinguistics* 12: 453–76.
Elliott, A. 2013. *Reinvention*. London: Routledge.
Elliott, A. and Urry, J. 2010. *Mobile Lives*. London: Routledge.
Ellwood-Clayton, B. 2005. Desire and Loathing in the Cyber-Philippines. In *The Inside Text: Social, Cultural and Design Perspectives on SMS*, ed. R. Harper, L. Palen and A. Taylor, pp. 195–219. Dordrecht: Springer.
Ensslin, A. 2011. 'What an un-wiki way of doing things': Wikipedia's Multilingual Policy and Metalinguistic Practice. *Journal of Language and Politics* 10: 535–61.
Everett, A. 2002. The Revolution Will Be Digitized: Afrocentricity and the Digital Public Sphere. *Social Text* 71: 125–46.
Fanon, F. [1952] 1986. *Black Skin, White Masks*. Trans. C.M. Markmann, forewords Z. Sadar and H.K. Bhabha. London: Pluto Press.
Fishman, J.A. 2001. *Can Threatened Languages Be Saved? Reversing Language Shift Revisited*. Clevedon: Multilingual Matters.
Fornäs, J., Klein, K., Ladendorf, M., Sundén, J. and Sveningsson, M. 2002. *Digital Borderlands: Cultural Studies of Identity and Interactivity on the Internet*. New York: Peter Lang.
Fortunati, L. 2005. Is Body-to-Body Communication Still the Prototype? *Information Society* 21: 53–61.
Foucault, M. [1977] 1980. Confessions of the Flesh. In: *Power/Knowledge: Selected Interviews and Other Writings 1972–1977*, ed. C. Gordon, trans. C. Gordon, L. Marshall, J. Mepham and K. Soper, pp. 194–228. New York: Pantheon Books.
Fuchs, C. 2014a. *Social Media: A Critical Introduction*. London: Sage.
Fuchs, C. 2014b. *Digital Labour and Karl Marx*. London: Routledge.
Gal, S. 1989. Language and Political Economy. *Annual Review of Anthropology* 18: 345–67.
Gallup, J. 2012. Precocious *Jouissance*: Roland Barthes, Amatory Maladjustment, and Emotion. *New Literary History* 43: 565–82.
Gardiner, M. 2004. Wild Publics and Grotesque Symposiums: Habermas and Bakhtin on Dialogue, Everyday Life and the Public Sphere. *Sociological Review* 52: 28–48.
Gatson, S.N. 2011. The Methods, Politics, and Ethics of Representation in Online Ethnography. In *Collecting and Interpreting Qualitative Materials*, ed. N.K. Denzin and Y.S. Lincoln, fourth edition, pp. 245–75. Thousand Oaks, CA: Sage.

Gauntlett, D. 2011. *Making is Connecting: The Social Meaning of Creativity, from DIY and Knitting to YouTube and Web 2.0.* Cambridge: Polity.

Gawne, L. and Vaughan, J. 2011. 'I can haz language play': The Construction of Language and Identity in LOLspeak. In *Proceedings of the 42nd Australian Linguistic Society Conference*, ed. M Ponsonnet, L. Dao and M Bowler, pp. 97–122. Canberra: ANU Research Repository.

Gee, J.P. 2005. Semiotic Social Spaces and Affinity Spaces: From the Age of Mythology to Today's Schools. In *Beyond Communities of Practice: Language, Power and Social Context*, ed. D. Barton and K. Tusting, pp. 214–32. Cambridge: Cambridge University Press.

Gershon, I. 2010. *The Breakup 2.0: Disconnecting over New Media.* Ithaca: Cornell University Press.

Gibson, J.J. 1979. *The Ecological Approach to Perception.* London: Houghton Mifflin.

Giddens, A. 1990. *The Consequences of Modernity.* Cambridge: Polity.

Giddens, A. 2011. *Runaway World: How Globalisation is Reshaping our Lives.* New edition. London: Profile Books.

Giles, J. 2005. Internet Encyclopaedias go Head to Head. *Nature* 438: 900–1

Gillen, J. 2013. Writing Edwardian Postcards. *Journal of Sociolinguistics* 17: 488–521.

Gillwald, A., Milek, A. and Stork, C. 2010. Towards Evidence-Based ICT Policy and Regulation: Gender Assessment of ICT Access and Usage in Africa. *ResearchICT.net. Vol. 1, Policy Paper 5.* Available at http://www.ictworks.org/sites/default/files/uploaded_pics/2009/Gender_Paper_Sept_2010.pdf.

Ginsburg, F. 2008. Re-Thinking the Digital Age. In *The Media and Social Theory*, ed. D. Hesmondhalgh and J. Toynbee, pp. 129–44. New York: Routledge.

Ginsburg, F. 2012. Disability in the Digital Age. In *Digital Anthropology*, ed. H.A. Horst and D. Miller, pp. 101–26. London/New York: Berg.

Glenn, P. 2003. *Laughter in Interaction.* Cambridge: Cambridge University Press.

Goffman, E. 1969. *The Presentation of Self in Everyday Life.* London: Penguin.

Goffman, E. 1974. *Frame Analysis: An Essay on the Organization of Experience.* New York: Harper and Row.

Goffman, E. 1979. *Gender Advertisements.* London: Macmillan.

Goffman, E. 1981. *Forms of Talk.* Philadelphia: University of Pennsylvania Press.

Goffman, E. 1983. The Interaction Order: American Sociological Association 1982, Presidential Address. *American Sociological Review* 48: 1–17.

Gong, H. and Yang, X. 2010. Digitized Parody: The Politics of *egao* in Contemporary China. *China Information* 24: 3–26.

Goody, J. 1987. *The Interface between the Written and the Oral.* Cambridge: Cambridge University Press.

Goriunova, O. 2012. *Art Platforms and Cultural Production on the Internet.* London: Routledge.

Gorny, E. 2009.The Virtual Persona as a Creative Genre on the Russian Internet. In *Control + Shift: Public and Private Usages of the Russian Internet*, ed. H. Schmidt, K. Teubener and N. Konradova, second revised edition, pp. 156–76. Norderstedt: Books on Demand.

Grant, T. 2010. Txt 4n6: Idiolect Free Authorship Analysis. In: *The Routledge Handbook of Forensic Linguistics*, ed. M. Couldhard and A. Johnson, pp. 508–22. London: Routledge.

Greenblatt, S. 1991. *Marvellous Possessions: The Wonder of the New World.* Oxford/ New York: Oxford University Press.

Grice, H.P. 1975. Logic and Conversation. In *Syntax and Semantics. Vol. 3*, ed. P. Cole and J. Morgan, pp. 41–58. New York: Academic Press.

Grinter, R.E. and Eldridge, M. 2001. y do tngrs luv 2 txt msg? In *Proceedings of the Seventh European Conference on Computer- Supported Cooperative Work ECSCW '01, Bonn, Germany*, ed. W. Prinz, M. Jarke, Y. Rogers, K. Schmidt and V. Wulf, pp. 219–38. Dordrecht: Kluwer Academic.

Grossman, L. 2006. You – Yes, You – Are TIME's Person of the Year. *Time Magazine*, 25 December. Available at http://content.time.com/time/magazine/article/0,9171,1570810,00.html.

Gumperz, J.J. 1964. Linguistic and Social Interaction in Two Communities. *American Anthropologist* 66: 137–53.

Gumperz, J.J. 1982. *Discourse Strategies*. Cambridge: Cambridge University Press.

Gumperz, J.J. 1992. Contextualization and Understanding. In *Rethinking Context: Language as an Interactive Phenomenon*, ed. A. Duranti and C. Goodwin, pp. 229–52. Cambridge: Cambridge University Press.

Gunkel, D. 2011. To Tell the Truth: The Internet and Emergent Epistemological Challenges in Social Research. In *The Handbook of Emergent Technologies in Social Research*, ed. S. Nagy Hesse-Biber, pp. 47–64. Oxford/New York: Oxford University Press.

Habermas, J. 1962. *Strukturwandel der Öffentlichkeit: Untersuchungen zu einer Kategorie der bürgerlichen Gesellschaft*. Neuwied: Luchterhand.

Hall, S. 1996. Who Needs 'Identity'? In *Questions of Cultural Identity*, ed. S. Hall and P. du Gay, pp. 1–17. Thousand Oaks, CA: Sage.

Hall, S. (ed.) 1997. *Representation*. London: Sage.

Halliday, M.A.K. 1978. *Language as Social Semiotic: The Social Interpretation of Language and Meaning*. Baltimore: University Park Press.

Hancock, J.T. 2007. Digital Deception: Why, When and How People Lie Online. In *The Oxford Handbook of Internet Psychology*, ed. A. Joinson, K. McKenna, T. Postmes and U.-D. Reips, pp. 289–301. Oxford/ New York: Oxford University Press.

Hanks, W. F. 1989. Text and Textuality. *Annual Review of Anthropology* 18: 95–127.

Haraway, D. [1985] 1991. A Cyborg Manifesto: Science, Technology, and Socialist-Feminism in the Late Twentieth Century. In *Simians, Cyborgs and Women: The Reinvention of Nature*, pp. 149–81. London: Routledge.

Harris, R. 1981. *The Language Myth*. London: Duckworth.

Harris, R. 2000. *Rethinking Writing*. London: Athlone.

Harris, Z.S. 1952. Discourse Analysis. *Language* 28: 1–30.

Havelock, E.A. 1982. *The Literate Revolution in Greece and its Cultural Consequences*. Princeton: Princeton University Press.

Heath, S.B. 1983. What No Bedtime Story Means: Narrative Skills at Home and School. *Language in Society* 11: 49–76.

Heller, M. and Duchêne, A. 2007. Discourses of Endangerment: Sociolinguistics, Globalization and the Social Order. In *Discourses of Endangerment*, ed. A. Duchêne and M. Heller, pp. 1–13. London/New York: Continuum.

Helsper, E. J. and Eynon, R. 2010. Digital Natives: Where is the Evidence? *British Educational Research Journal* 36: 503–20.

Henn-Memmesheimer, B. and Eggers, E. 2010. Inszenierung, Etablierung und Auflösung: Karriere einer grammatischen Konstruktion im Chat zwischen 2000 und 2010. *NET.WORX* 57: 5–34.

Herring, S.C. 2003. Gender and Power in Online Communication. In *The Handbook of Language and Gender*, ed. J. Holmes and M. Meyerhoff, pp. 202–28. Oxford: Blackwell.

Herring, S.C. 2012. Grammar and Electronic Communication. In *The Encyclopedia of Applied Linguistics*, ed. C. Chapelle (online publication). Hoboken, NJ: Wiley-Blackwell.

Herring, S.C. and Zelenkauskaite, A. 2009. Symbolic Capital in a Virtual Heterosexual Market: Abbreviation and Insertion in Italian iTV SMS. *Written Communication* 26: 5–31.

Hill, J.H. 1995. The Voices of Don Gabriel: Responsibility and Self in a Modern Mexicano Narrative. In *The Dialogic Emergence of Culture*, ed. D. Tedlock and B. Mannheim, pp. 97–147. Chicago: University of Illinois Press.

Hill, J.H. 2008. *The Everyday Language of White Racism*. Oxford: Wiley-Blackwell.

Hinnenkamp, V. 2008. Deutsch, Doyc or Doitsch? Chatters as Languagers – The Case of a German-Turkish Chat Room. *International Journal of Multilingualism* 5: 253–75.
Hjorth, L., Burgess, J. and Richardson, I. (eds.) 2012. *Studying Mobile Media: Cultural Technologies, Mobile Communication, and the iPhone*. London: Routledge.
Hodges, A. 2011. *The 'War on Terror Narrative': Discourse and Intertextuality in the Construction and Contestation of Sociopolitical Reality*. Oxford/New York: Oxford University Press.
Horkheimer, M. and Adorno, T.W. [1944] 2002. *Dialectic of Enlightenment: Philosophical Fragments*. Ed. G. Schmid Noer, trans. E. Jephcott. Stanford: Stanford University Press.
Horst, H.A. and Miller, D. 2006. *The Cell Phone: An Anthropology of Communication*. Oxford/New York: Berg.
Horst, H.A. and Miller, D. (eds.) 2012. *Digital Anthropology*. London/New York: Berg.
Humboldt, W. von [1836] 1998. *Über die Verschiedenheit des menschlichen Sprachbaus und ihren Einfluss auf die geistige Entwicklung des Menschengeschlechts*. Ed. D. Di Cesare. Paderborn: Schöningh.
Humboldt, W. von [1827] 1994. *Über den Dualis*. In *Über die Sprache: Reden vor der Akademie*, ed. J. Trabant, pp. 143–69. Tübingen: Francke.
Hutchby, I. 2001. *Conversation and Technology: From the Phone to the Internet*. Cambridge: Polity.
Hymes, D. [1974] 2004. Breakthrough into Performance. In *'In vain I tried to tell you': Essays in Native American Ethnopoetics*, pp. 79–141. Philadelphia: University of Pennsylvania Press.
Hymes, D. 1996. *Ethnography, Linguistics, Narrative Inequality: Towards a Theory of Voice*. New York: Taylor and Francis.
Iedema, R. 2003. Multimodality, Resemiotization: Extending the Analysis of Discourse as Multi-Semiotic Practice. *Visual Communication* 2: 29–57.
Independent Commission on International Development Issues 1980. *North–South: A Programme for Survival*. London: Pan Books.
Ishii, K. 2004. Internet Use Via Mobile Phone in Japan. *Telecommunications Policy* 28: 43–57.
Ito, M. and Okabe, D. 2005. Technosocial Situations: Emergent Structurings of Mobile E-Mail Use. In *Personal, Portable, Pedestrian: Mobile Phones in Japanese Life*, ed. M. Ito, D. Okabe and M. Matsuda, pp. 257–75. Cambridge, MA: MIT Press.
Ito, M., Baumer, S., Bittani, M., boyd, d., Cody, R., Horst, H., Lange, P., Mahendran, D., Martínez, K., Pascoe, C.J., Perkel, D., Robinson, L., Sims, C. and Tripp, L. 2010. *Hanging Out, Messing Around, and Geeking Out: Kids Living and Learning with New Media*. Cambridge, MA/London: MIT Press.
ITU 2013. *Measuring the Information Society*. Geneva: International Communications Union.
Ivkovic, D. and Lotherington, H. 2009. Multilingualism in Cyberspace: Conceptualising the Virtual Linguistic Landscape. *International Journal of Multilingualism* 6: 17–36.
Jacquemet, M. 2005. Transidiomatic Practices: Language and Power in the Age of Globalization. *Language and Communication* 25: 257–77.
Jakobson, R. 1960. Linguistics and Poetics. In *Style in Language*, ed. T.A. Seboek, pp. 350–77. New York: Technology Press.
James, C. 1986. *Other Passports: Poems 1958–1985*. London: Jonathan Cape.
Jargon File 2004. Version 4.4.8. Available at http://www.catb.org/jargon.
Jenkins, H. 2006. *Fans, Bloggers, and Gamers: Exploring Participatory Culture*. New York: New York University Press.
Joas, H. 1996. *The Creativity of Action*. Chicago: Chicago University Press.
Johnson, D., Pejovic, V., Belding, E. and Van Stam, G. 2011. Traffic Characterization and Internet Usage in Rural Africa. In *20th International World Wide Web Conference*

(WWW11), Hyderabad, India, 28 March–1 April 2011. Available at http://www.cs.ucsb.edu/~ebelding/txt/www11.pdf.

Jones, G. and Schieffelin, B.B. 2009. Enquoting Voices, Accomplishing Talk: Uses of *be + like* in Instant Messaging. *Language and Communication* 29: 77–113.

Jones, R. (ed.) 2012. *Discourse and Creativity*. Harlow: Pearson.

Jones, R. and Hafner, C.A. 2012. *Understanding Digital Literacies: A Practical Introduction*. London: Routledge.

Jonsson, C. and Muhonen, A. 2012. 'Jalla ronaldo eat them! ♥': Indexing Superdiversity through Polylingual Digital Literary Practices on Facebook. Paper presented at Sociolinguistic Symposium 19, 21–4 August, Berlin.

Jørgensen, J.N. 2008. Polylingual Languaging Around and Among Children and Adolescents. *International Journal of Multilingualism* 5: 161–76.

Kapidzic, S., and Herring, S.C. 2011. Gender, Communication, and Self-Presentation in Teen Chatrooms Revisited: Have Patterns Changed? *Journal of Computer-Mediated Communication* 17: 39–59.

Kaschula, R.H. and Mostert, A. 2009. The Influence of Cellular Phone 'Speak' on isiXhosa Rules of Communication. *Stellenbosch Papers in Linguistics PLUS* 37: 69–88.

Kataoka, K. 1997. Affect and Letter-Writing: Unconventional Conventions in Casual Writing by Young Japanese Women. *Language in Society* 26: 103–36.

Katz, J.E. and Aakhus, M. (eds.) 2002. *Perpetual Contact: Mobile Communication, Private Talk and Public Performance*. Cambridge: Cambridge University Press.

Kellner, D. 2009. Barack Obama and Celebrity Spectacle. *International Journal of Communication* 3: 715–41.

Kelly-Homes, H. 2005. *Advertising in Multilingual Communication*. Basingstoke: Palgrave Macmillan.

Kemp, N. 2010. Texting versus txtng: Reading and Writing Text Messages, and Links with Other Linguistic Skills. *Writing Systems Research* 2: 53–71.

Kirkpatrick, D. 2010. *The Facebook Effect*. London: Virgin Books.

Knuttila, L. 2011. User Unknown: 4chan, Anonymity and Contingency. *First Monday* 16, 3 October. Available at http://firstmonday.org/ojs/index.php/fm/article/view/3665/3055.

Kornei, A. 2013. Digital Language Death. *PLOS ONE* 8: 11.

Kramsch, C. 2009. *The Multilingual Subject: What Foreign Language Learners Say about Their Experiences and Why it Matters*. Oxford: Oxford University Press.

Kress, G. 2000. *Early Spelling: Between Convention and Creativity*. London: Routledge.

Kress, G. 2010. *Multimodality: A Social Semiotic Approach to Contemporary Communication*. London: Routledge.

Kristeva, J. [1969] 1980. Word, Dialogue and Novel. In *Desire in Language: A Semiotic Approach to Literature and Art*, pp. 64–91. New York: Columbia University Press.

Kristeva, J. [1974] 1980. The Ethics of Linguistics. In *Desire in Language: A Semiotic Approach to Literature and Art*, pp. 23–35. New York: Columbia University Press.

Kulick, D. 2003. No. *Language and Communication* 23: 139–51.

Kuntsman, A. and Stein, R.L. 2011. Digital Suspicion, Politics, and the Middle East. *Critical Inquiry* (online feature on Arab Spring). Available at http://criticalinquiry.uchicago.edu/digital_suspicion_politics_and_the_middle_east.

Kuss, D.J. and Griffiths, M.D. 2011. Online Social Networking and Addiction: A Review of the Psychological Literature. *International Journal of Environmental Research and Public Health* 8: 3528–52.

Labov, W. 1972. *Sociolinguistic Patterns*. Philadelphia: University of Pennsylvania Press.

Labov, W. 2001. *Principles of Linguistic Change: Social Factors*. Oxford: Blackwell.

Labov, W. 2010. *Principles of Linguistic Change: Cognitive and Cultural Factors*. Oxford: Wiley-Blackwell.

Lamoureaux, S. 2011. *Message in a Mobile: Mixed-Messages, Tales of Missing and Mobile Communities at the University of Khartoum*. Bamenda: Langaa.
Langford, W. 1997. 'Bunnikins, I love you snugly in your warren': Voices from Subterranean Cultures of Love. In *Language and Desire: Encoding Sex, Romance and Intimacy*, ed. K. Harvey and C. Shalom, pp. 170–85. London: Routledge.
Lapidos, J. 2013. Yes we Can to Yes we Scan. *New York Times*, 18 July. Available at http://takingnote.blogs.nytimes.com/2013/07/18/yes-we-can-to-yes-we-scan/?_r=0.
Larmer, B. 2011. When an Internet Joke is not Just a Joke. *New York Times*, 26 October. Available at http://www.nytimes.com/2011/10/30/magazine/the-dangerous-politics-of-internet-humor-in-china.html?pagewanted=all.
Latour, B. 1999. *Pandora's Hope: Essays on the Reality of Science Studies*. Cambridge, MA: Harvard University Press.
Lave, J. and Wenger, E. 1991. *Situated Learning: Legitimate Peripheral Participation*. Cambridge: Cambridge University Press.
Law, J. and Urry, J. 2004. Enacting the Social. *Economy and Society* 33: 390–410.
Lawton, A. and Eagle, H. 2004. *Words in Revolution: Russian Futurist Manifestoes 1912–1928*. Washington, DC: New Academia.
Le Page, R.B. and Tabouret-Keller, A. 1985. *Acts of Identity: Creole-Based Approaches to Language and Ethnicity*. Cambridge: Cambridge University Press.
Lekhanya, T. 2013. *Cell Phone Practices and Language Choices Among Basotho Migrants in De Doorns, Western Cape, South Africa*. Unpublished Honours thesis, University of Cape Town.
Lemphane, P. and Prinsloo, M. 2013. Children's Digital Literacy Practices in Unequal South African Settings. *Tilburg Papers in Culture Studies* 60. Available at https://www.tilburguniversity.edu/research/institutes-and-research-groups/babylon/tpcs.
Lenhart, A., Ling, R., Campbell, S. and Purcell, K. 2010. *Teens and Mobile Phones*. Pew Internet & American Life Project. Available at http://www.pewinternet.org/Reports/2010/Teens-and-Mobile-Phones.aspx.
Leppänen, S. and Peuronen, S. 2013. Multilingualism and the Internet. In *The Encyclopedia of Applied Linguistics*, ed. C. Chapelle (online publication). Hoboken, NJ: Wiley-Blackwell.
Lewis, M.P., Simons, G.F. and Fennig, C.D. (eds.) 2013. *Ethnologue: Languages of the World*. Seventeenth edition. Dallas: SIL International. Online version available at http://www.ethnologue.com.
Lexander, K.V. 2011. Texting and African Language Literacy. *New Media Society* 13: 427–43.
Li Wei 2011. Moment Analysis and Translanguaging Space: Discursive Construction of Identities by Multilingual Chinese Youth in Britain. *Journal of Pragmatics* 43: 1222–35.
Lieberson, S. 2000. *A Matter of Taste: How Names, Fashions and Culture Change*. New Haven/London: Yale University Press.
Lillis, T. 2013. *The Sociolinguistics of Writing*. Edinburgh: Edinburgh University Press.
Lin, A.M.Y. and Tong, A. 2007. Text-Messaging Cultures of College Girls in Hong Kong: SMS as Resources for Achieving Intimacy and Gift-Exchange with Multiple Functions. *Continuum: Journal of Media and Cultural Studies* 21: 303–15.
Ling, R. and Baron, N.S. 2007. Text Messaging and IM Linguistic Comparison of American College Data. *Journal of Language and Social Psychology* 26: 291–8.
Ling, R. and Horst, H.A. 2011. Mobile Communication in the Global South. *New Media Society* 13: 363–74.
Ling, R. and Yttri, B. 2002. Hyper-Coordination via Mobile Phones in Norway. In *Perpetual Contact: Mobile Communication, Private Talk, Public Performance*, ed. J. Katz and M. Aakhus, pp. 139–69. Cambridge: Cambridge University Press.
Link, P. and Xiao Qiang 2013. From Grass-Mud Equestrians to Rights-Conscious Citizens: Language and Thought on the Chinese Internet. In *Restless China*, ed. P. Link and P.G. Pickowicz, pp. 83–106. Plymouth: Rowman and Littlefield.

Lovink, G. and Tkacz, N. (eds.) 2011. *Critical Point of View: A Wikipedia Reader*. Amsterdam: Institute for Network Cultures.
Lunt, P. and Livingstone, S. 2013. Media Studies' Fascination with the Concept of the Public Sphere: Critical Reflections and Emerging Debates. *Media, Culture & Society* 35: 87–96.
Makoni, S. and Pennycook, A. (eds.) 2007. *Disinventing and Reconstituting Languages*.
Manning, P.K. 1996. Dramaturgy, Politics and the Axial Media Event. *Sociological Quarterly* 37: 261–28.
Manning, P.K. 2005. *Freud and American Sociology*. Cambridge: Polity.
Marcus, G.E. 1995. Ethnography in/of the World System: The Emergence of Multi-Sited Ethnography. *Annual Review of Anthropology* 24: 95–117.
Marcuse, H. 1964. *One-Dimensional Man: Studies in the Ideology of Advanced Industrial Society*. Boston: Beacon.
Marinetti, F.M. [1909] 2009. The Founding and Manifesto of Futurism. In *Futurism: An Anthology*, ed. L. Rainey, C. Poggi and L.Wittman, pp. 49–53. New Haven/London: Yale University Press.
Marwick, A. and boyd, d. 2011. I Tweet Honestly, I Tweet Passionately: Twitter Users, Context Collapse, and the Imagined Audience. *New Media & Society* 13: 114–33.
Maybin, J. and Swann, J. 2007. Everyday Creativity in Language: Textuality, Contextuality, and Critique. *Applied Linguistics* 28: 497–517.
Mazzerella, W. 2010. Beautiful Balloon: The Digital Divide and the Charisma of New Media in India. *American Ethnologist* 37: 783–804.
Mbembe, A. 2004. Aesthetics of Superfluidity. *Public Culture* 16: 373–405.
McIntosh, J. 2010. Mobile Phones and Mipoho's Prophecy: The Powers and Dangers of Flying Language. *American Ethnologist* 37: 337–53.
McKean, E. 2002. L33t-sp34k. *Verbatim: The Language Quarterly* 27: 13–14.
McLuhan, M. 1964. *Understanding Media: The Extensions of Man*. Mentor: New York.
Mendoza, N. 2011. A Tale of Two Worlds: Apocalypse, 4Chan, WikiLeaks and the Silent Protocol Wars. *Radical Philosophy* 166: 1–8.
Meng, B. 2011. From Steamed Bun to Grass Mud Horse: E Gao as Alternative Political Discourse on the Chinese Internet. *Global Media and Communication* 7: 33–51.
Mesthrie, R., Swann, J. Deumert, A. and Leap, W. 2009. *Introducing Sociolinguistics*. Edinburgh: Edinburgh University Press.
Meyrowitz, J. 1985. *No Sense of Place: The Impact of Electronic Media on Social Behaviour*. Oxford: Oxford University Press.
Miller, D. 2011. *Tales from Facebook*. Cambridge: Polity.
Miller, L. 2011. Subversive Script and Novel Graphs in Japanese Girls' Culture, *Language and Communication* 31: 16–26.
Miller, R. 1967. *The Japanese Language*. Tokyo: Charles E. Tuttle.
Mufwene, S. 2008. *Language Evolution: Contact, Competition, and Change*. London: Continuum.
Murphey, L.L. and Priebe, A.E. 2011. "My co-wife can borrow my mobile phone!': Gendered Geographies of Cell Phone Usage and Significance for Rural Kenyans. *Gender, Technology and Development* 15: 1–23.
Myers, G. 2010. *The Discourse of Blogs and Wikis*. London/New York: Continuum.
Nahon, K. and Hemsley, J. 2013. *Going Viral*. Cambridge: Polity.
Nakassis, C.V. 2012. Brand, Citationality and Performativity. *American Anthropologist* 114: 624–38.
Nkomo, D. and Khumalo, L. 2012. Embracing the Mobile Phone Technology: Its Social and Linguistic Impact with Special Reference to Zimbabwean Ndebele. *African Identities* 10: 143–53.
Ochs, E. 1990. Indexicality and Socialization. In *Cultural Psychology*, ed. J.W. Stigler, R.A. Shweder and G. Herdt, pp. 287–308. Cambridge: Cambridge University Press.

Oldenburg, R. 1989. *The Great Good Place: Cafes, Coffee Shops, Community Centers, Beauty Parlors, General Stores, Bars, Hangouts, and How They Get You Through the Day*. New York: Paragon House.

Olson, D. 1994. *The World on Paper: The Conceptual and Cognitive Implications of Writing and Reading*. Cambridge: Cambridge University Press.

Ong, W. 1982. *Orality and Literacy: The Technologizing of the Word*. New York: Methuen.

Oreglia, E. and Kaye, J. 2012. A Gift from the City: Mobile Phones in Rural China. In *Proceedings of the ACM 2012 Conference on Computer Supported Cooperative Work* (CSCW '12), pp. 137–46. New York: ACM.

Otsuji, E. and Pennycook, A. 2010. Metrolingualism: Fixity, Fluidity and Language in Flux. *International Journal of Multilingualism* 7: 240–54.

Page, R. 2010. Re-Examining Narrativity: Small Stories in Status Updates. *Text & Talk* 30: 423–44.

Palfrey, J. and Gasser, U. (eds.) 2011. *Born Digital: Understanding the First Generation of Digital Natives*. London: Basic Books.

Palfreyman, D. and Al Khahil, M. 2007. 'A funky language for teens to use': Representing Gulf Arabic in Instant Messaging. In *The Multilingual Internet: Language, Culture and Communication Online*, ed. B. Danet and S.C. Herring, pp. 43–63. Oxford/New York: Oxford University Press.

Panciera, K., Halfaker, A. and Terveen, L. 2009. Wikipedians Are Born Not Made: A Study of Power Editors on Wikipedia. In *GROUP '09: Proceedings of the ACM 2009 International Conference on Supporting Group Work*. Available at http://grouplens.org/system/files/Group09WikipediansPanciera.pdf.

Paolillo, J.C. 2011. 'Conversational' Codeswitching on Usenet and Internet Relay Chat. *Language @ Internet 8*. Available at http://www.languageatinternet.org/articles/2011/Paolillo.

Pennycook, A. 2007. *Global Englishes and Transcultural Flows*. London/New York: Routledge.

Pennycook, A. 2012. *Language and Mobility: Unexpected Places*. Bristol: Multilingual Matters.

Pentzold, C. 2009. Fixing the Floating Gap: The Online Encyclopaedia Wikipedia as a Global Memory Place. *Memory Studies* 2: 255–72.

Perley, B.C. 2012. Zombie Linguistics: Experts, Endangered Languages and the Curse of Undead Voices. *Anthropological Forum* 22: 133–49.

Perloff, M. 1986. *The Futurist Movement: Avant-Garde, Avant Guerre, and the Language of Rupture*. Chicago: University of Chicago Press.

Peters, J.D. 1999. *Speaking into the Air: A History of the Idea of Communication*. Chicago: University of Chicago Press.

Pfaff, J. 2010. A Mobile Phone: Mobility, Materiality and Everyday Swahili Trading Practices. *Cultural Geographies* 17: 341–57.

Phillips, W. 2011. LOLing at Tragedy: Facebook Trolls, Memorial Pages and Resistance to Grief Online. *First Monday*, 5 December. Available at http://firstmonday.org/ojs/index.php/fm/article/view/3168/3115.

Phillips, W. 2013. The House that Fox Built: Anonymous, Spectacle, and Cycles of Amplification. *Television & New Media* 14: 494–509.

Pieterse, J.N. 1992. *White on Black: Images of Africa and Blacks in Western Popular Culture*. New Haven: Yale University Press.

Piller, I. and Takahashi, K. 2006. A Passion for English: Desire and the Language Market. In *Bilingual Minds: Emotional Experience, Expression and Representation*, ed. A. Pavlenko, pp. 59–83. Clevedon: Multilingual Matters.

Pimienta, D., Prado, D. and Blanco, A. 2009. *Twelve Years of Measuring Linguistic Diversity on the Internet: Balance and Perspectives*. Paris: UNESCO.

Plester, B., Wood, C. and Bell, V. 2008. Txt msg n school literacy: Does Texting and Knowledge of Text Abbreviations Adversely Affect Children's Literacy Attainment? *Literacy* 42: 137–44.

Porter, G., Hamshire, K., Abane, A., Munthali, A., Robson, E., Mashiri, M. and Tanle, A. 2012. Youth, Mobility and Mobile Phones in Africa: Findings from a Three-Country Study. *Information Technology for Development* 18: 145–62.

Pound, L. 1923. Spelling-Manipulation and Present-Day Advertising. *Dialect Notes* 5: 226–32.

Pound, L. 1925. The Kraze for 'K'. *American Speech* 1: 43–4.

Prado, D. 2012. Language Presence in the Real World and in Cyber Space. In *NET.LANG: Towards the Multilingual Cyberspace*, ed. L. Vannini and H. Le Crosnier, pp. 35–52. Caen: C&F éditions.

Prensky, M. 2001. Digital Natives, Digital Immigrants. *On the Horizon* 9: 1–6.

Prensky, M. 2011. Digital Wisdom and Homo Sapiens Digital. In *Deconstructing Digital Natives: Young People, Technology and the New Literacies*, ed. M. Thomas, pp. 15–29. London: Routledge.

Prinsloo, M. 2005. The New Literacies as Placed Resources. *Perspectives in Education* 24: 1–12.

Putnam, R. 2000. *Bowling Alone: The Collapse and Revival of American Community*. New York: Simon & Schuster.

Qiu, J.L. 2009. *Working-Class Network Society: Communication Technology and the Information Have-Less in Urban China*. Cambridge, MA: MIT Press.

Ragnedda, M. and Muschert, G.W. (eds.) 2013. *The Digital Divide: The Internet and Social Inequality in International Perspective*. London: Routledge.

Rampton, B. 1995. *Crossing: Language and Ethnicity Among Adolescents*. London: Longman.

Rampton, B. 2009. Interaction Ritual and Not Just Artful Performance in Crossing and Stylization. *Language in Society* 38: 149–76.

Rampton, B. 2011. Style Contrasts, Migration and Social Class. *Journal of Pragmatics* 43: 1236–50.

Rampton, B. and Charalambous, C. 2012. Crossing. In *The Routledge Handbook of Multilingualism*, ed. M. Martin-Jones, A. Blackledge and A. Creese, pp. 482–98. London: Routledge.

Ratele, K. 2004. Kinky Politics. In *Re-Thinking Sexualities in Africa*, ed. S. Arnfred, pp. 139–57. Uppsala: Nordic Africa Institute.

Read, A.W. 1963. The First Stage in the History of 'O.K.'. *American Speech* 38: 5–27.

Rettie, R. 2009. Mobile Phone Communication: Extending Goffman to Mediated Interaction. *Sociology* 43: 421–38.

Rheingold, H. 1993. *Virtual Communities: Homesteading the Electronic Frontier*. Reading, MA: Addison-Wesley.

Rheingold, H. 2002. *Smart Mobs: The Next Social Revolution*. Cambridge, MA: Perseus Books.

Rivon, V. 2012. The Use of Facebook by the Eton in Cameroon. In *NET.LANG: Towards the Multilingual Cyberspace*, ed. L. Vannini and H. Le Crosnier, pp. 161–8. Caen: C&F éditions.

Robinson, K.K. and Crenshaw, E.M. 2010. Reevaluating the Global Digital Divide: Socio-Demographic and Conflict Barriers to the Internet Revolution. *Sociological Inquiry* 80: 34–62.

Rymes, B. 2012. Recontextualizing YouTube: From Macro-Micro to Mass-Mediated Communicative Repertoires. *Anthropology and Education Quarterly* 43: 214–27.

SalahEldeen, H.M. and Nelson, M.L. 2012. Losing My Revolution: How Many Resources Shared on Social Media Have Been Lost? In *Proceedings of the Second International Conference on Theory and Practice of Digital Libraries*, TPDL '12, pp. 125–37. Available at http://arxiv.org/pdf/1209.3026.pdf.

Salih, S. 2002. *Judith Butler*. London: Routledge.

Sapir, E. 1917. Do We Need a 'Superorganic'? *American Anthropologist* 19: 411–47.

Sapir, E. 1921. *Language: An Introduction to the Study of Speech*. New York: Harcourt.
Sapir, E. 1927. Speech as a Personality Trait. *American Journal of Sociology* 32: 892–905.
Sapir, E. [1938] 1968. Why Cultural Anthropology Needs the Psychiatrist. In *Selected Writings of Edward Sapir in Language, Culture and Personality*, ed. D.G. Mandelbaum, pp. 569–77. Berkeley/Los Angeles: University of California Press.
Schechner, R. 2002. *Performance Studies: An Introduction*. London: Routledge.
Schieffelin, B.B. 2000. Introducing Kaluli Literacy: A Chronology of Influences. In *Regimes of Language*, ed. P. Kroskrity, pp. 293–327. Santa Fe: School of American Research Press
Schnoebelen, T. 2012. Do You Smile with Your Nose? Stylistic Variation in Twitter Emoticons. *University of Pennsylvania Working Papers in Linguistics* 18. Available at http://repository.upenn.edu/cgi/viewcontent.cgi?article=1242&context=pwpl.
Schoon, A.J. 2012. Dragging Young People Down the Drain: The Mobile Phone, Gossip Mobile Website *Outoilet* and the Creation of a Mobile Ghetto. *Critical Arts: South-North Cultural and Media Studies* 26: 690–706.
Schultze, U. and Mason, R.O. 2012. Studying Cyborgs: Re-Examining Internet Studies as Human Subjects Research. *Journal of Information Technology* 27: 301–12.
Schwartz, M. 2008. The Trolls Among Us. *New York Times*, 3 August. Available at http://www.nytimes.com/2008/08/03/magazine/03trollst.html?pagewanted=all&_r=0.
Scollon, R. and Scollon, S.B. 2003. *Discourses in Place: Language and the Material World*. London: Routledge.
Scribner, S. and Cole, M. 1981. *The Psychology of Literacy*. Cambridge, MA: Harvard University Press.
Seargeant, P. and Tagg, C. (eds.) 2014. *The Language of Social Media: Identity and Community on the Internet*. Basingstoke: Palgrave Macmillan.
Sebba, M. 2007. *Spelling and Society*. Cambridge: Cambridge University Press.
Sey, A. 2011. 'We use it different, different': Making Sense of Trends in Mobile Phone Use in Ghana. *New Media & Society* 13: 375–90.
Sherzer, J. 2002. *Speech Play and Verbal Art*. Austin: University of Texas Press.
Shields, R. 2003. *The Virtual*. London: Routledge.
Shifman, L. 2013. *Memes in Digital Culture*. Cambridge, MA: MIT Press.
Shohamy, E. and Gorter, D. (eds.) 2009. *Linguistic Landscape: Expanding the Scenery*. London: Routledge.
Siegel, J. 2008. In Praise of the Cafeteria Principle: Language Mixing in Hawai'i Creole. In *Roots of Creole Structures: Weighing the Contributions of Substrates and Superstrates*, ed. S. Michaelis, pp. 59–82. Amsterdam: John Benjamins.
Silk, M. 2010. The Language of Greek Lyric Poetry. In *A Companion to the Ancient Greek Language*, ed. E.J. Bakker, pp. 424–40. Oxford: Wiley-Blackwell.
Silva, C. 2011. Writing in Portuguese Chats:) A new wrtng systm? *Written Language and Literacy* 14: 143–56.
Silverstein, M. 2003. Indexical Order and the Dialectics of Sociolinguistic Life. *Language and Communication* 23: 193–229.
Silverstein, M. and Urban, G. 1996. The Natural History of Discourse. In *Natural Histories of Discourse*, ed. M. Silverstein and G. Urban, pp. 1–20. Chicago: University of Chicago Press.
Simmel, G. 1949. The Sociology of Sociability (trans. E.C. Hughes). *American Journal of Sociology* 55: 254–61.
Sivapragasam, N. and Kang, J. 2011. The Future of the Public Payphone: Findings from a Study of Telecom Use at the Bottom of the Pyramid in South and Southeast Asia. *Information Technologies and International Development* 7: 33–44.
Smith, D.J. 2006. Cell Phones, Social Inequality, and Contemporary Culture in Nigeria. *Canadian Journal of African Studies* 40: 496–523.

Smith, N. 2004. *Chomsky: Ideas and Ideals.* Second edition. Cambridge: Cambridge University Press.

Snell, J. 2010. From Sociolinguistic Variation to Socially Strategic Stylisation. *Journal of Sociolinguistics* 14: 630–56.

Soffer, O. 2012. Liquid Language? On the Personalization of Discourse in the Digital Era. *New Media Society* 14: 1092–110.

Sperlich, W. 2005. Will Cyberforums Save Endangered Languages? A Niuean Case Study. *International Journal of the Sociology of Language* 172: 51–77.

Spitulnik, D. 1996. The Social Circulation of Media Discourse and the Mediation of Communities. *Journal of Linguistic Anthropology* 6: 161–87.

Squires, L. 2010. Enregistering Internet Language. *Language in Society* 39: 457–92.

Stevenson, N. 2010. New Media, Popular Culture and Social Theory. In *The Routledge Companion to Social Theory*, ed. A. Elliott, pp. 156–72. London: Routledge.

Storch, A. 2011. *Secret Manipulations: Language and Context in Africa.* Oxford: Oxford University Press.

Street, B. 1984. *Literacy in Theory and Practice.* Cambridge: Cambridge University Press.

Street, B. (ed.) 1993. *Cross-Cultural Approaches to Literacy.* Cambridge: Cambridge University Press.

Stroud, C. forthcoming. Afterword: Turbulent Deflections. In *Language, Literacy, and Diversity: Moving Words*, ed. C. Stroud and M. Prinsloo. London: Routledge.

Sturrock, J. 1986. Jamboree. *London Review of Books* 8: 13–14.

Sturrock, J. 1993. *Structuralism.* Oxford: Blackwell.

Su, H.-Y. 2007. The Multilingual and Multiorthographic Taiwan-Based Internet: Creative Uses of Writing Systems on College-Affiliated BBS. In *The Multilingual Internet: Language, Culture and Communication Online*, ed. B. Danet and S.C. Herring, pp. 64–86. Oxford: Oxford University Press.

Sutherland, M. 2012. Populism and Spectacle. *Cultural Studies* 26: 330–45.

Swann, J., Pope, R. and Carter, R. (eds.) 2011. *Creativity in Language and Literature: The State of the Art.* Basingstoke: Palgrave Macmillan.

Tacchi, J., Kitner, K.R. and Crawford, K. 2012. Meaningful Mobility. *Feminist Media Studies* 12: 528–37.

Tagg, C. 2009. A *Corpus Linguistics Study of SMS Text Messaging.* Unpublished PhD thesis, University of Birmingham.

Tagg, C. 2012 *Discourse of Text Messaging: Analysis of SMS Communication.* London/New York: Continuum.

Tagg, C. and Seargeant, P. 2012. Writing Systems at Play in Thai–English Online Interactions. *Writing Systems Research* 4: 195–213.

Tagg, C. and Seargeant, P. 2014. Audience Design and Language Choice in the Construction and Maintenance of Translocal Communities on Social Networking Sites. In *The Language of Social Media: Identity and Community on the Internet*, ed. P. Seargeant and C. Tagg, pp. 161–85. Basingstoke: Palgrave Macmillan.

Tagliamonte, S.A. and Denis, D. 2008. Linguistic Ruin? Lol! Instant Messaging and Teen Language. *American Speech* 83: 3–34.

Takahashi, K. 2013. *Language Learning, Gender and Desire.* Clevedon: Multilingual Matters.

Takahashi, T. 2010. MySpace or Mixi? Japanese Engagement with SNS (Social Networking Sites) in the Global Age. *New Media & Society* 12: 453–75

Tannen, D. 2007. *Talking Voices: Repetition, Dialogue and Imagery in Conversational Discourse.* Second edition. Cambridge: Cambridge University Press.

Tapscott, D. 1998. *Growing Up Digital: The Rise of the Net Generation.* New York: McGraw-Hill.

Taylor, A. and Harper, R. 2002. Age-Old Practices in the 'New World': A Study of Gift-Giving Between Teenage Mobile Phone Users. In *Proceedings of Conference on Human*

Factors in Computing Systems CHI '02, Minneapolis, MN, pp. 439–46. New York: ACM Press.

Taylor, J.R. 2012. *The Mental Corpus: How Language is Represented in the Mind.* Oxford: Oxford University Press.

Temple, O., Berry, L., Jack, B., Laudiwana, O., Lawes, N., Maipe, E., Salle, D., Sion, F. and Winnia, X. 2011. *Tok Ples in Texting and Social Networking: PNG 2010. The Impact of Mobile Phones and SMS Technology on the Indigenous Languages of Papua New Guinea.* Waigani: UPNG Press.

Thomason, S. 2000. *Language Contact: An Introduction.* Edinburgh: Edinburgh University Press.

Thorne, S.L. 2012. Gaming Writing: Supervernaculars, Stylization, and Semiotic Remediation. In *Technology across Writing Contexts and Tasks*, ed. G. Kessler, A. Oskoz, and I. Elola, pp. 297–316. San Marcos, Texas: CALICO Monographs.

Thurlow, C. 2003. Generation TxT? The Sociolinguistics of Young People's Text-Messaging. *Discourse Analysis Online.* Available at http://extra.shu.ac.uk/daol/articles/v1/n1/a3/thurlow2002003–01.html.

Thurlow, C. 2006. From Statistical Panic to Moral Panic: The Metadiscursive Construction and Popular Exaggeration of New Media Language in the Print Media. *Journal of Computer-Mediated Communication* 11: 667–701.

Thurlow, C. 2012. Determined Creativity: Language Play, Vernacular Literacy and New Media Discourse. In *Discourse and Creativity*, ed. R. Jones, pp. 169–90. London: Pearson.

Thurlow, C. and Mroczek, K. (eds.) 2011. *Digital Discourse: Language in the New Media.* Oxford: Oxford University Press.

Thurlow, C. and Poff, M. 2013. Text Messaging. In *Handbook of the Pragmatics of CMC*, ed. S.C. Herring, D. Stein and T. Virtanen, pp. 163–90. Berlin/New York: De Gruyter.

Toma, C.L. and Hancock, J.T. 2012. What Lies Beneath: The Linguistic Traces of Deception in Online Dating Profiles. *Journal of Communication* 62: 78–97.

Toolan, M. 2012. Literary Creativity. In *Linguistic Creativity*, ed. R. Jones, pp. 15–34. Harlow: Pearson.

Tseliga, T. 2007. 'It's All Greeklish to Me!': Linguistic and Sociocultural Perspectives on Roman-Alphabeted Greek in Asynchronous Computer-Mediated Communication. In *The Multilingual Internet: Language, Culture and Communication Online*, ed. B. Danet and S.C. Herring, pp. 116–41. Oxford: Oxford University Press.

Turkle, S. [1984] 2005. *The Second Self: Computers and the Human Spirit.* Twentieth anniversary edition. Cambridge, MA/London: MIT Press.

Turkle, S. 1995. *Life on the Screen: Identity in the Age of the Internet.* New York: Simon & Schuster.

Turkle, S. 2007. *Evocative Objects: Things We Think With.* Cambridge, MA: MIT Press.

Turkle, S. 2008. Always-On/Always-On-You: The Tethered Self. In *Handbook of Mobile Communication Studies*, ed. J.E. Katz, pp. 121–37. Cambridge, MA: MIT Press.

Turkle, S. 2011. *Alone Together: Why We Expect More from Technology and Less from Each Other.* New York: Basic Books.

UNDP 2013. *The Rise of the South: Human Progress in a Diverse World.* Available at http://hdr.undp.org/en/content/human-development-report-2013.

UNESCO 2003. *Language Vitality and Endangerment*, comp. by M. Brenzinger et al. Paris: UNESCO.

Ureta, S. 2008. Mobilizing Poverty? Mobile Phone Use and Everyday Spatial Mobility Among Low-Income Families in Santiago, Chile. *Information Society* 24: 83–91.

Urry, J. 2000. *Sociology beyond Societies: Mobilities for the Twenty-First Century.* London: Routledge.

Urry, J. 2007. *Mobilities.* Cambridge: Polity.

Uy-Tioco, C. 2007. Overseas Filipino Workers and Text Messaging: Reinventing Transnational Mothering. *Continuum: Journal of Media & Cultural Studies* 21: 253–65.
Vaisman, C.L. 2014. Beautiful Script, Cute Spelling and Glamorous Words: Doing Girlhood through Language Playfulness on Israeli Blogs. *Language and Communication* 34: 69–80.
Van Blerk, L. 2010. *Identity Work and Community Work through Sociolinguistic Rituals on Facebook*. Unpublished Honours research thesis, University of Cape Town.
Van Dijk, J.A.G.M. 2013. A Theory of the Digital Divide. In *The Digital Divide: The Internet and Social Inequality in International Perspective*, ed. M. Ragnedda and G.W. Muschert, pp. 29–51. London: Routledge.
Van Zoonen, L. 2013. From Identity to Identification: Fixating the Fragmented Self. *Media, Culture & Society* 35: 44–51.
Vandekerckhove, R. and Nobels, J. 2010. Code Eclecticism: Linguistic Variation and Code Alternation in the Chat Language of Flemish Teenagers. *Journal of Sociolinguistics* 14: 657–77.
Velten, H.R. 2012. Performativity and Performance. In *Travelling Concepts for the Study of Culture*, ed. B. Neumann and A. Nünning, pp. 249–66. Berlin: De Gruyter.
Vertovec, S. 2007. Super-Diversity and its Implications. *Ethnic and Racial Studies* 29: 1024–54.
Vice, S. 1997. *Introducing Bakhtin*. Manchester: Manchester University Press.
Vigouroux, C. 2011. Magic Marketing: Performing Grassroots Literacy. *Diversities* 13: 53–69.
Waldorf, L. 2012. White Noise: Hearing the Disaster. *Journal of Human Rights Practice* 4: 469–74.
Wall, M. 2009. Africa on YouTube: Musicians, Tourists, Missionaries and Aid Workers. *International Communication Gazette* 71: 393–407.
Wallis, C. 2011. Mobile Phones without Guarantees: The Promises of Technology and the Contingencies of Culture. *New Media & Society* 13: 471–85.
Walton, M. 2011. Mobilizing African Publics. *Information Technologies and International Development* 7: 47–50.
Warschauer, M. and Mtuchniak, T. 2010. New Technology and Digital Worlds: Analyzing Evidence of Equity in Access, Use and Outcomes. *Review of Research in Education* 34: 179–225.
Wasserman, H. 2011. Mobile Phones, Popular Media, and Everyday African Democracy: Transmissions and Transgressions. *Popular Communications* 9: 146–58.
Weilenmann, A. and Larsson, C. 2002. Local Use and Sharing of Mobile Phones. In *Wireless World: Social and Interactional Aspects of the Mobile Age*, ed. B. Brown, N. Green and R. Harper, pp. 92–107. London: Springer.
Weinreich, U., Labov, W. and Herzog, M.I. 1968. Empirical Foundations for a Theory of Language Change. In *Directions for Historical Linguistics*, ed. W. P. Lehmann, pp. 95–105. Austin: University of Texas Press.
Wesch, M. 2009. YouTube and You: Experiences of Self-Awareness in the Context Collapse of the Recording Webcam. *Explorations in Media Ecology* 8: 19–34.
Williams, Q. and Stroud, C. 2010. Performing Rap Ciphas in Late-Modern Cape Town: Extreme Locality and Multilingual Citizenship. *Afrika Focus* 23: 39–59.
Williams, R. 1974. *Television: Technology and Cultural Form*. London: Collins.
Williams, R. [1976] 1983 *Keywords*. Second edition. Oxford/New York: Oxford University Press.
Winter, D. 2011. *MXing in Afrikaans*. Unpublished Honours research thesis, University of Cape Town.
Wood, C., Meachem, S., Bowyer, S., Jackson, E., Tarczynski-Bowles, M.L. and Plester, B. 2011. A Longitudinal Study of Children's Text Messaging and Literacy Development. *British Journal of Psychology* 102: 431–42.

Woolard, K.A. 2004. Code-Switching. In *A Companion to Linguistic Anthropology*, ed. A. Duranti, pp. 73–94. Oxford: Blackwell.

Yang, G. 2003. The Internet and the Rise of a Transnational Chinese Cultural Sphere. *Media, Culture & Society* 25: 469–90.

Yang, H. 2009. *The Code-Switching Patterns of Mandarin–English Bilinguals in Computer-Mediated Communication*. Unpublished Honours research thesis, University of Cape Town.

Yoshiki, M. and Kodama, S. 2012. Measuring Linguistic Diversity on the Net. In *NET.LANG: Towards the Multilingual Cyber*space, ed. L. Vannini and H. Le Crosnier, pp. 35–52. Caen: C&F éditions.

Young, R. 1985/6. Back to Bakhtin. *Cultural Critique* 2: 71–92.

Yu, H. 2007. Blogging Everyday Life in Chinese Internet Culture. *Asian Studies Review* 31: 423–33.

Zainudeen, A., Iqbal, T. and Samarajiva, R. 2010. Who's Got the Phone? Gender and the Use of the Telephone at the Bottom of the Pyramid. *New Media & Society* 12: 549–66.

Zappavigna, M. 2014. CoffeeTweets: Bonding Around the Bean of Twitter. In *The Language of Social Media: Identity and Community on the Internet*, ed. P. Seargeant and C. Tagg, pp. 139–60. Basingstoke: Palgrave Macmillan.

Zentella, A. 1997. *Growing up Bilingual: Puerto Rican Children in New York*. Oxford: Blackwell.

Index